Keys to the Marine Invertebrates of Puget Sound, the San Juan Archipelago, and Adjacent Regions

Keys to the Marine Invertebrates of Puget Sound, the San Juan Archipelago, and Adjacent Regions

By EUGENE N. KOZLOFF

With contributions by several specialists

UNIVERSITY OF WASHINGTON PRESS

SEATTLE AND LONDON

Printed in the United States of America

Library of Congress Cataloging in Publication Data

Kozloff, Eugene N
Keys to the marine invertebrates of Puget Sound, the San Juan Archipelago, and adjacent regions.

Includes bibliographies.
1. Marine invertebrates--Puget Sound--Identification. 2. Marine invertebrates--Washington (State)--San Juan Islands--Identification. I. Title.
QL138.K68 592'.09'2632 74-12027
ISBN 0-295-95347-0

PREFACE

Over a period of more than twenty years, many zoologists associated with Friday Harbor Laboratories contributed to a collection of keys for identification of marine invertebrates of the San Juan Archipelago and adjacent regions. This material was reproduced from time to time by mimeograph, so its circulation was limited. The unfinished quality of many portions probably discouraged submission of criticisms, suggestions, and additions.

Although this published edition of the keys is by no means definitive, the treatment of some groups is substantially complete, and the manual will now be readily accessible to those who need it. Thus it should be useful to a large number of students and professional biologists working in this region, and we may hope that rigorous testing of the keys will, within a few years, lead to many improvements.

In the course of my work on this manual, I have been aided by many colleagues. Those who prepared keys and checklists for particular groups of invertebrates, or who assembled information on which keys are based, are acknowledged in connection with the sections to which they contributed. I thank all of them here, without mentioning them individually.

To Paul Illg, whose knowledge of our invertebrate fauna surely surpasses that of any other zoologist, I offer especially warm thanks. He allowed me to use his notes on earlier versions of keys, some of them developed by students under his direction, his records of collections made in this area, and his extensive personal library. He also gave me good advice whenever I asked for it. To another colleague, Robert Fernald, director of Friday Harbor Laboratories during most of my term of service there, I am grateful for the support he gave to the development of a reference collection and to production of mimeographed editions of keys, as well as for his efforts in promoting systematic studies by a number of students and specialists.

Dennis Willows, the present director of Friday Harbor Laboratories, provided extensive and valued logistic support during a period when I was able to devote full time to this work. Much credit is also due to Estelle Johnson, administrative secretary, and to Carol De Boer for keeping things moving and for their attention to odd problems that arose while I was enjoying my reverie of subchelate gnathopods and homogomph falcigerous setae.

The list of acknowledgments must also include Dixy Lee Ray and Emery Swan. Their encouragement of work on our fauna, begun more than twenty years ago, led some of their students and associates to prepare keys and to make extensive collections that later figured in published systematic studies. Two of the earlier keys that served faithfully for a number of years and that were helpful in the preparation of this manual are those of Harry Fritchman and Richard Snyder. More recent keys and reports that have been especially useful are those of Anne Hurst, Goshusp Anklesaria John, Lynda Richards, Kenneth Boss, Tai Soo Park, and Scott Herrmann. The influence of several visiting professors on faunistic and systematic studies at Friday Harbor was important in the development of much of the material upon which the present keys are based.

During the earlier phases of my own work, Gretchen Lambert compiled checklists, combed the literature for descriptive details, perfected bibliographic citations, and corresponded with specialists. Without her assistance and interest, I might not have been able to make progress rapid enough to encourage me to continue. In later phases of the program, Carl Nyblade was particularly helpful in connection with my attempts to construct the complex keys to shelled gastropods and anomuran decapods. His knowledge of these groups and of recent literature concerned with them saved me from making some embarrassing mistakes.

In the preparation of this type of publication, an author or editor must rely on very accurate typing and strict adherence to format. The major part of the manuscript was typed by Joy Strickland. Her devotion to the task, as well as her good humor and patience in dealing with an author who had to change his mind frequently, are sincerely appreciated. I also wish to thank Patricia Schoenwolf, Pamela Peterman, and my dauther, Rae, for typing certain sections. The careful work of Elizabeth Vance in preparing typescripts of earlier versions of some of the keys made revisions much easier than they would have been otherwise.

For several groups of animals covered by this manual, I received suggestions, corrections, and additions from a number of colleagues. Some were kind enough to read and correct extensive sections. I will simply list those who made direct contributions in one way or another to keys and checklists concerned with various classes or phyla of invertebrates. Porifera: Gerald Bakus; Cnidaria: Lemuel Fraser; Nemertea: Pamela Roe and Fumio Iwata; Mollusca: Paul Raymore and H. E. Scheidt; Annelida: Karl Banse, Katherine Hobson, Sarah Woodin, Frederick Nichols, Eugene Burreson, Colin Hermans, and Bertil Swedmark; Sipuncula and Echiura: Mary Rice; Crustacea: Placidus Reischman, Raymond Bauer and Richard Brusca; Bryozoa: June Ross; Echinodermata: Arthur Fontaine and Norman Engstrom.

A number of others assisted in the solution of various problems, brought in material that was useful for making comparisons of species, looked after the reference collection of invertebrates, or helped in some other way toward the production of this manual. In mentioning those I can remember, I take the risk of inadvertently overlooking some. In any case, I thank Dana Read, Joyce Lewin, Gail Irvine, Cleave Vandersluys, Alan Kohn, Stephen Bloom, Charles Moffett, Steven Brocco, and David Galloway.

Ralph Smith and James Carlton, editors of the new edition of *Light's Manual*, covering invertebrates of the coast of central California, generously permitted me to study the manuscript, and thus to become informed about nomenclatorial changes and recently published systematic studies that I would otherwise have missed. Certain of the keys prepared for *Light's Manual* helped me in constructing keys to the same groups as they are represented in our fauna.

A professional leave from the University of Washington made it possible for me to concentrate for several months on the task of completing this work. I am most grateful to the University for this opportunity. Grants to Friday Harbor Laboratories by the National Science Foundation, used in part to support systematic studies, are also acknowledged. Finally, I wish to thank my wife, Anne, for helping me look for mistakes in the manuscript and for contributing to the preparation of the index.

In spite of the great amount of assistance I have received in connection with this manual, it is inevitable that there will be errors, inconsistencies, ambiguities, and important omissions. No one is more painfully aware than I of the difficulties associated with compiling a set of keys that are at the same time reasonably complete, practical, up-to-date in nomenclature and terminology, and foolproof in the hands of a nonspecialist. I will value every suggestion and correction that is sent to me, and will study it carefully with a view to incorporating its substance into future revisions.

EUGENE N. KOZLOFF

Department of Zoology
University of Washington
Seattle

SOURCES OF ILLUSTRATIONS

Many of the figures illustrating terms used in the keys have been copied, or redrawn with only slight modification, from works of other authors. The sources of these are gratefully acknowledged.

Porifera: G. J. Bakus, in *Journal of Zoology*, London, 1966.
Hydrozoa: D. V. Naumov, in *Opredeliteli po Faune SSSR*, 1960.
Kinorhyncha: R. P. Higgins, in *Transactions of the American Microscopical Society*, 1960.
Mollusca: W. H. Dall, in *Bulletin of the United States National Museum*, 1921; Z. A. Filatova and V. I. Zatsepin, in *Opredelitel' Fauny i Flory Severnikh Moreĭ SSSR* (editor: N. S. Gaevskaia͡), 1948; T. Habe, in *Publications from the Akkeshi Marine Biological Station*, 1955; F. M. MacFarland, in *Memoirs of the California Academy of Sciences*, 1966.
Polychaeta: M. H. Pettibone, in *Bulletin of the United States National Museum*, 1963, and *Smithsonian Contributions to Zoology*, 1971; P. V. Ushakov, in *Opredeliteli po Faune SSSR*, 1955.
Crustacea: J. L. Barnard, in *Bulletin of the United States National Museum*, 1969, and *Smithsonian Contributions to Zoology*, 1972; F. A. Chace, Jr., in *Smithsonian Contributions to Zoology*, 1972; J. C. McCain, in *Bulletin of the United States National Museum*, 1968; W. L. Schmitt, in *University of California Publications in Zoology*, 1921; B. E. Stafford, in *First Annual Report of the Laguna Marine Laboratory*, 1912.
Bryozoa: R. C. Osburn, in *Allan Hancock Foundation Pacific Expeditions*, 1950-53.
Ophiuroidea: A. M. D'i͡akonov, in *Opredeliteli po Faune SSSR*, 1967.
Ascidiacea: W. G. Van Name, in *Bulletin of the American Museum of Natural History*, 1945.

CONTENTS

Keys to the Marine Invertebrates of Puget Sound, the San Juan Archipelago, and Adjacent Regions

INTRODUCTION

This manual has been developed with practicality in mind. Wherever possible, the keys are based on obvious characters and terminology has been simplified. This approach, however, is not usually suitable for construction of keys that will differentiate families, orders, and other higher taxa in addition to genera and species. Ideally, then, each key should be followed by a checklist in which the genera and species are organized according to families, orders, and other higher taxa. Annotations concerning color, size, geographic distribution, habitat preferences, breeding season, and other aspects of biology would also be helpful. Unfortunately, addition of such material would greatly increase the cost of this manual, as well as delay its publication. The need for functional keys to most groups of marine invertebrates of this region subdues the passion for perfection.

Geographic Area Covered

Although the keys and checklists produced at Friday Harbor have traditionally been concerned with the fauna of the San Juan Archipelago, they have been expanded to cover Puget Sound and the Strait of Georgia, as well as the open coast of Washington and southern Vancouver Island. In fact, they will function rather well as far south as Coos Bay on the Oregon coast. In a few cases, where animals known to occur on the southern Oregon coast are not actually included in the keys, notes alluding to their presence in that region--and sometimes a note about their distinctive characteristics--are included in the appropriate section.

Structure of the Keys

All of the keys are strictly dichotomous. Each numbered couplet consists of two choices. Begin with the first couplet--1a and 1b--moving next to whichever couplet is indicated following the better of the two choices. If neither choice seems to agree quite perfectly with the specimen in hand, it will be a good idea to follow both possibilities. Hopefully, one will soon find that the specimen seems to match the characteristics of several successive choices of one lineage but not of the other. A very real danger in using any key is that a species not known to occur in the region will be misidentified. An effort has been made to minimize this possibility by giving, whenever possible, more than one pair of contrasting characteristics in each couplet, and sometimes including information on habitat preferences of the species involved. In some cases, a note inserted within or following a couplet comments on a problem of systematics or indicates that a similar species may be expected in our region.

Checklists and Other Substitutions for Keys

Certain groups of invertebrates in our area are poorly known. Publication of keys to such groups would very likely lead to misidentification of species that have not previously been found in this region, or that have not even been described. If only a few species have been reported for a particular group, these are listed, and the bibliography will lead to descriptions and illustrations of all or most of them. However, if the group is large and one that only a specialist can be expected to deal with--as the Nematoda--the only material that can logically be provided is a bibliography of papers known to be relevant or at least predicted to be useful.

In the case of certain groups that have been extensively studied in our area, the bibliographies will lead to published works in which there are keys, detailed descriptions, and illustrations. This approach has been taken with the calanoid copepods and Euphausiacea, both of which consist of many planktonic species not likely to be dealt with by users of this manual. As long as keys to calanoids and euphausiaceans are available to those who need them, it would seem inappropriate to reproduce them in their entirety in this work.

Illustrations and Glossary

It is assumed that anyone using this manual will have access to a textbook of zoology or invertebrate zoology, so that recapitulation of the characteristics of phyla, classes, and most other higher taxa may be omitted. However, although the terminology of the keys has been simplified as much as possible, it is inevitable that many of the descriptive terms employed will be unusual. These terms are defined in the glossary, and certain characteristics are illustrated in connection with the keys to which they apply.

Bibliographies

The references provided in connection with the keys and checklists are, as a rule, restricted to citations of works judged to be relevant to the identification of invertebrates of this region. However, works that have long been superseded or whose content has been assimilated into recent studies may be omitted, unless they have special value because of illustrations, locality records, or further bibliographic data they contain. The bibliographic treatment accorded one group may be very different from that given to another, depending on the extent to which these groups, as represented in our area, have been studied. In some cases, references not certain to be useful in this region are cited because they may be helpful in recognizing genera or species that are to be expected, even if they have not yet been reported.

Some Useful General References

As the bibliographies appended to the separate sections of this manual include, as a rule, only monographs or papers that are concerned with specific groups of invertebrates, a few particularly useful general references will be mentioned here.

Smith, R. I., & J. T. Carlton. 1974. *Light's Manual: Intertidal Invertebrates of the Central California Coast.* Berkeley and Los Angeles: University of California Press.

Ricketts, E. F., & J. Calvin. 1968. *Between Pacific Tides.* 4th edition, revised by J. W. Hedgpeth. Stanford, California: Stanford University Press. 614 pp.

Kozloff, E. N. 1973. *Seashore Life of Puget Sound, the Strait of Georgia, and the San Juan Archipelago.* Seattle & London: University of Washington Press. 282 pp.

Although *Light's Manual* is devoted largely to the invertebrates of the coast of central California, its bibliographies are very extensive, including almost all important contributions dealing with systematics of invertebrates of the west coast of the United States and Canada. It is also valuable because it cites many papers concerned with ecology, behavior, physiology, and other aspects of the biology of Pacific coast invertebrates, and is a good guide to encyclopedic works concerned with the morphology and classification of various groups.

Between Pacific Tides, which deals in considerable detail with natural history and ecology of invertebrates of central California, and to some extent with those of areas farther south and farther north, also has an extensive bibliography. However, the new edition of *Light's Manual* is more nearly complete and up to date, especially with respect to works concerned with systematics.

Seashore Life is written in nontechnical language and will be helpful in providing an overview of the invertebrates, seaweeds, and fishes that are associated with various habitats, including floating docks, in the Pacific Northwest. This little book identifies, at least to genus (and usually to species), almost two hundred invertebrates, and provides photographs and drawings of some others that are not identified but are representative of the groups to which they belong.

PHYLUM PORIFERA

1a Spicules calcareous (dissolved by hydrochloric acid) Class Calcarea

1b Spicules (these are absent in 3 of our species) not dissolved by hydrochloric acid 2

2a Spicules basically 6-rayed, but they may be modified by reduction or loss of certain rays, by branching of rays, etc.; at least many of the spicules fused to form a firm lattice; strictly subtidal Class Hyalospongia (Hexactinellida)

2b Spicules (these are absent in 3 of our species) not basically 6-

rayed, though they are represented by many forms, some of which do have rays; spicules not generally fused to form a firm lattice; in-includes most of the intertidal and subtidal species of our area
Class Demospongiae

CLASS CALCAREA

The two species in the key are the only ones that are at all abundant in our area. For any which do not seem to fit, the paper of de Laubenfels (1932) should be referred to, but identification beyond genus will probably not be possible. Moreover, identification even to genus may require the preparation of sections. Burton (1963) has drastically reorganized the classification of calcareous sponges, and his work must be referred to by anyone concerned with the taxonomy of this difficult group.

1a Vase-shaped (up to about 5 cm tall) and with a conspicuous osculum, attached by the base, and generally single; common on floats, especially during the summer, but also found intertidally and subtidally
Scypha ?raphanus (Schmidt, 1862)

1b Consisting of a mass of branching (and occasionally anastomosing) tubes about 1 mm in diameter; taken intertidally and subtidally, and sometimes becoming established in large tanks supplied with running sea water, but not often common on floats *Leucosolenia* sp.

CLASS HYALOSPONGIA (HEXACTINELLIDA)

The most common species in our area is *Aphrocallistes vastus* Schulze, 1886 (*A. whiteavesianus* Lambe, 1893) (Aphrocallistidae). It forms urnlike growths, often more than 20 cm high, and occurs at depths of about 40 m or more. Other species that have been reported from the northern part of this general region are *Rhodocalyptus dawsoni* (Lambe, 1893) and *Staurocalyptus dowlingii* (Lambe, 1894) (Rossellidae), and *Chonelasma calyx* Schulze, 1886 (Euretidae). They have not been found in Puget Sound or the San Juan Archipelago. The first two of these are illustrated in the respective papers of Lambe. The work of Ijima (1927) presents a useful overview of the hexactinellid sponges, and thus may be helpful in recognizing genera encountered in the course of dredging in deep water, especially off the coast.

CLASS DEMOSPONGIAE

The literature pertinent to this group of sponges in our area has been admirably reviewed by Bakus (1966). Bakus also presented a detailed analysis of 23 species belonging to the order Poecilosclerida, and attempted to interpret descriptions of earlier authors, especially Lambe and de Laubenfels (see bibliography for citations of those papers concerned directly with the fauna in this general region).

The key provided here is inevitably incomplete, for several reasons. The fauna of poecilosclerida sponges of the outer coast is not as well known as that of inland waters, especially in the San Juan Archipelago. Moreover, Demospongiae of other orders have not been accorded the same kind of treatment given by Bakus to the Poecilosclerida. Finally, some of the species described or reported by earlier workers have not been found, or at least have not been restudied to the extent that they can be differentiated from closely related species that can be recognized with certainty.

The key takes advantage of color, texture, and general form, as well as spicule types. In the case of the poecilosceleridan sponges, the identifications can be confirmed by referring to Bakus' monograph.

1a Without spicules (other than a few from other sponges that may have been incorporated into the colony); thin encrustations, usually slippery to the touch — 2

1b With spicules; ranging from thin encrustations to thick masses, sometimes branching, and usually with a feltlike surface or a surface that is smooth but not slippery — 4

2a Without spongin fibers; light brown, dark brown, pale gray, or lavender gray — *Halisarca* sp.
See de Laubenfels, 1948 for a review of this genus, as well as of *Aplysilla*.

2b With spongin fibers (these support little cones scattered over the surface) — 3

3a Color nearly colorless to rose or red — *Aplysilla glacialis* (Merejkowsky, 1880)

3b Color deep purple (material of this color exudes from the sponge when it is handled) — *Aplysilla polyraphis* de Laubenfels, 1930

4a Boring in shells of molluscs and barnacles, the portions of the colony showing at the surface generally constituting small yellow patches — *Cliona celata* Grant, 1826

4b Not boring in shells — 5

5a Colony shaped like a slender funnel — *Stylissa stipitata* de Laubenfels, 1961

5b Colony not shaped like a funnel — 6

6a Colony erect, with a number of gnarled, fingerlike branches; surface rather smooth; color generally brownish yellow — *Syringella amphispicula* de Laubenfels, 1961

6b Colony not branched; surface smooth, feltlike or with a thick plush of large spicules; color variable — 7

7a Colony globular or hemispherical, several cm thick, with a thick external plush of large spicules that originate deep within the colony and radiate outward (the spicules may be more than 2 cm long) — 8

7b Colony usually thin and encrusting, but if massive, without a thick external plush of large spicules — 9

8a With 1 or more conspicuous oscula; without a number of smooth, nipplelike protuberances; color usually whitish beige or whitish gray — *Tetilla* sp.

8b Without obvious oscula; with a number of smooth, nipplelike protuberances; color usually yellowish, brightest on the protuberances — *Polymastia pachymastia* de Laubenfels, 1932

9a Colonies irregularly ovoid, up to several cm thick, firm and rubbery, with a smooth surface (without obvious protruding spicules); frequently inhabited by a hermit crab (the sponge grows over a gastropod shell in which a hermit crab is living, and eventually the shell is dissolved) — *Suberites ficus* (Johnston, 1842)

9b Colonies ranging from thin encrustations to thick masses, generally with a feltlike surface owing to small projecting spicules; not regularly associated with hermit crabs — 10

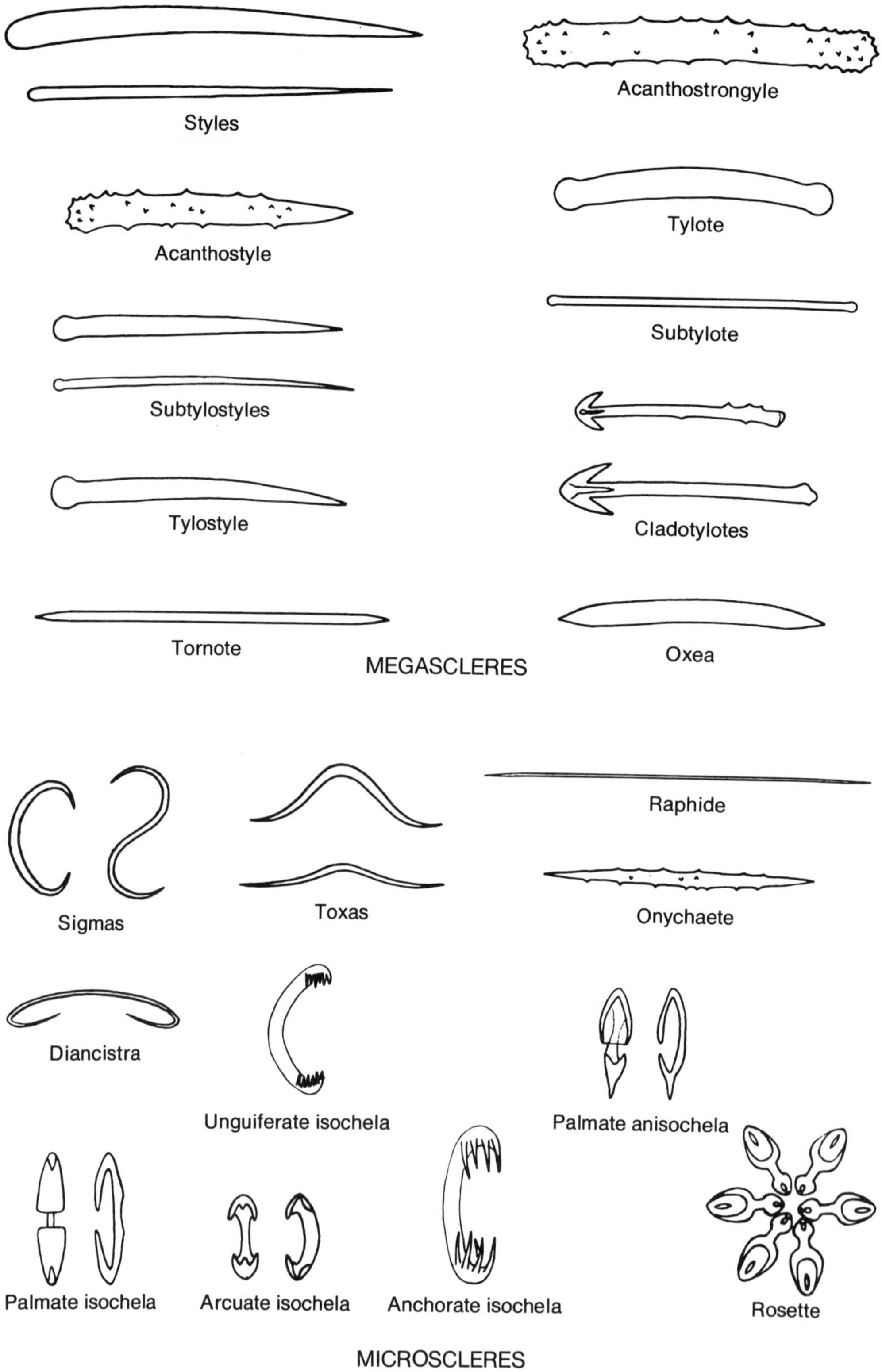
Styles
Acanthostrongyle
Acanthostyle
Tylote
Subtylote
Subtylostyles
Tylostyle
Cladotylotes
Tornote
Oxea
MEGASCLERES
Raphide
Sigmas
Toxas
Onychaete
Diancistra
Unguiferate isochela
Palmate anisochela
Palmate isochela
Arcuate isochela
Anchorate isochela
Rosette
MICROSCLERES

10a Color bright red or orange-red; colonies up to about 1 cm thick, but usually much thinner 11

10b Color whitish, drab, tan, some shade of yellow or brown, gray, green, violet, pink, etc. (but not definitely red or orange-red); colonies ranging from very thin to several cm thick 15

11a Acanthostyles and/or acanthostrongyles represented among the megascleres; colony up to about 1 cm thick 12

11b Neither acanthostyles nor acanthostrongyles represented among the megascleres; colony generally less than 5 mm thick (*Plocamia karykina* may be up to about 7 mm thick) 13

12a With both acanthostyles and acanthostrongyles; surface often roughened, but without distinct craterlike depressions
Plocamilla illgi Bakus, 1966

12b With acanthostyles, but without acanthostrongyles; surface pitted with small craterlike depressions
Microciona microjoanna de Laubenfels, 1930

13a Megascleres represented chiefly by styles or subtylostyles, but tylotes lacking; colony not exuding mucus when injured 14

13b Megascleres represented by tylotes as well as by styles and subtylostyles; colony often exuding mucus when injured
Plocamia karykina de Laubenfels, 1927

14a Megascleres all or mostly styles (sometimes a few subtylostyles are present); microscleres represented by palmate isochelas
Esperiopsis originalis de Laubenfels, 1930

14b Megascleres all or mostly subtylostyles (sometimes a few styles and tylostyles are present); microscleres represented by toxas
Ophlitaspongia pennata (Lambe, 1895)
In the variety *californiana* de Laubenfels, 1932, the length of the large subtylostyles ranges from about 210 to 286 μ, whereas in the typical form their length ranges from about 284 to 917 μ.

15a Megascleres represented only by oxeas; microscleres usually absent (except in *Sigmadocia*, which has sigmas) 16

15b Megascleres represented by spicules other than oxeas; microscleres almost always present (except in *Podotuberculum*) 19

16a Some sigmas present, in addition to oxeas; forming large brainlike masses; oscula not over 2 mm in diameter; subtidal
Sigmadocia ?edaphus (de Laubenfels, 1930)

16b Only oxeas present; ranging from thin encrustations to masses 3 or more cm thick; oscula may be several mm in diameter (*Haliclona*) or small; includes some of the most common intertidal and subtidal sponges, and the 2 species predominating on floats (*Haliclona permollis*, *Halichondria panicea*) 17

17a Oscula conspicuous, up to about 4 mm in diameter, usually situated on prominent volcanolike or tubular elevations; spicules arranged in triangular networks; color variable, generally grayish or violet
Haliclona permollis (Bowerbank, 1866)

17b Oscula small, not conspicuous; spicules not arranged in a network; color usually drab, yellowish, tan, grayish green, or green, but not violet 18

18a Forming encrustations up to 2 mm thick; texture almost stony; color drab *Xestospongia vanilla* (de Laubenfels, 1930)

18b Usually forming irregular masses up to 2 or 3 cm thick (sometimes more); texture soft or rubbery, but not stony; color usually yellowish or tan, sometimes grayish green or green (this species often has a strong odor, which may be disagreeable to some persons) *Halichondria panicea* (Pallas, 1766)

19a Megascleres represented by at least some acanthostyles (although the spines on these may be few) 20

19b Megascleres not represented by acanthostyles 26

20a Megascleres including some typical styles; colonies up to about 2 cm thick 21

20b Megascleres not including any typical styles, though other smooth megascleres (subtylostyles, tylostyles, tornotes) may be present; colonies ranging from thin encrustations to thick masses 23

21a Microscleres absent; color orange-brown to chocolate; with prominent tubercles that are a little brighter in color than the rest of the colony *Podotuberculum hoffmanni* Bakus, 1966

21b Microscleres represented by arcuate isochelas and sometime by sigmas; color yellow-brown to cinnamon; surface smooth or rough, but not really tuberculate 22

22a Subtylotes present; sigmas regularly present; yellow-brown to cinnamon *Lissodendoryx firma* (Lambe, 1895)

22b Subtylotes present; sigmas, if present, are probably contaminants; pinkish *Anthoarcuata graceae* Bakus, 1966

23a Subtylostyles present; colonies seldom more than 1 mm thick, smooth and mucoid to the touch; color light to dark cinnamon brown; subtidal *Hymedesanisochela rayae* Bakus, 1966

23b Subtylostyles absent; thickness of colonies variable, but texture not usually mucoid; color variable; includes some species found intertidally 24

24a Subtylotes present, tornotes absent; forming colonies up to about 15 cm thick *Iophon pattersoni* (Bowerbank, 1866)

24b Subtylotes absent, tornotes present; thickness of colonies variable 25

25a Microscleres represented by sigmas and arcuate isochelas; thickness variable, but may reach several cm; oscula up to about 4 mm in diameter; color generally nearly yellow or yellow-brown; intertidal and subtidal, and regularly found on (but not restricted to) valves of live scallops *Myxilla incrustans* (Esper, 1805-14)

25b Microscleres represented only by arcuate isochelas; thickness not often over 2 mm; oscula much less than 1 mm in diameter; color persimmon to yellow-brown, often with a slightly greenish tint; strictly subtidal, and rare *Hymendectyon lyoni* Bakus, 1966

26a Megascleres including typical styles 27

26b Megascleres not including typical styles 36

27a Either subtylotes or tornotes present 28

27b Neither subtylotes nor tornotes present 32

28a Subtylotes present, tornotes absent 29

28b Subtylotes absent, tornotes present 30

29a Some cladotylotes present; forming irregular masses or columns which may be taller than wide; oscula up to about 7 mm in diameter; color bronze *Acarnus erithacus* de Laubenfels, 1927

29b Cladotylotes absent; colonies up to about 1.5 cm thick; oscula up to about 3 mm in diameter; color orange-brown *Lissodendoryx* sp. (close to *L. kyma* de Laubenfels, 1930)

30a Sigmas and arcuate isochelas present; onychaetes absent; oscula up to about 4 mm in diameter; color generally nearly yellow or yellow-brown; intertidal and subtidal, and regularly found on (but not restricted to) valves of live scallops *Myxilla incrustans* (Esper, 1805-14)

30b Sigmas and arcuate isochelas absent; onychaetes present; oscula not obvious; color pinkish; strictly subtidal 31

31a Onychaetes all of approximately same size *Tedania fragilis* Lambe, 1895

31b Onychaetes falling into two distinct size ranges *Tedania gurjanovae* Koltun, 1958

32a Sigmas absent, toxas present *Axocielita hartmani* Simpson, 1966

32b Sigmas present, toxas absent 33

33a Raphides present, neither isochelas nor anisochelas present; colonies almost hemispherical, about 5 cm in diameter; color amber to olive brown *Biemna rhadia* de Laubenfels, 1930

33b Raphides absent, either isochelas or anisochelas present; colonies spreading, not likely to be almost hemispherical; color variable, but including yellowish, beige, light brown, violet, etc. 35

34a Rosettes of larger anisochelas present; oscula up to about 4 mm in diameter; color light brown to violet; commonly found on (but not restricted to) valves of live scallops *Mycale adhaerens* (Lambe, 1894)

34b Rosettes of larger anisochelas absent (but smaller rosettes may be present); oscula rarely more than 2 mm in diameter; color generally yellow (sometimes with a rosy tint), yellowish beige, yellowish brown, or reddish brown, but not violet; not likely to be found on valves of live scallops 35

35a Sigmas 14 to 25 μ; color ranging from yellow (sometimes with a rosy tint) to yellowish beige *Mycale richardsoni* Bakus, 1966

35b Sigmas 31 to 59 μ; color yellowish brown to reddish brown *Mycale hispida* (Lambe, 1894)

36a Subtylotes, anchorate isochelas, and unguiferate isochelas present, sigmas absent; color gray; when placed in alcohol, a black, flaky precipitate is given off *Stelotrochota hartmani* Bakus, 1966

36b Subtylotes, anchorate isochelas, and unguiferate isochelas absent (but other types of isochelas may be present), sigmas present; color usually some shade of yellow or brown; not giving off a black, flaky precipitate when placed in alcohol 37

37a Isochelas absent, but diancistras present (these may look like isochelas in profile view); up to only about 1 mm thick *Zygherpe hyaloderma* de Laubenfels, 1932

37b Isochelas present, diancistras absent; usually forming thick or even massive colonies 38

38a Raphides present (but may not be numerous) *Mycale lingua* (Bowerbank, 1866)

38b Raphides absent *Mycale psila* (de Laubenfels, 1930)

Bakus, G. J. 1966. Marine poeciloscleridan sponges of the San Juan Archipelago, Washington. J. Zool., London, 149:415-531.

Burton, M. 1963. A Revision of the Classification of Calcareous Sponges; with a Catalogue of the Specimens in the British Museum (Natural History). London: British Museum (Natural History). v + 693 pp.

Ijima, I. 1927. The Hexactinellida of the Siboga Expedition. Siboga - Expeditie, 6 (livr. 106). viii + 383 pp.

Lambe, L. M. 1893. On some sponges from the Pacific coast of Canada and Behring Sea. Trans. Royal Soc. Canada, Sect. IV, 10 (1892):67-78.

------. 1894. Sponges from the Pacific coast of Canada. Trans. Royal Soc. Canada, Sect. IV, 11 (1893):25-43.
------. 1895. Sponges from the western coast of North America. Trans. Royal Soc. Canada, Sect. IV, 12 (1894):113-38.
------. 1900. Catalog of the recent marine sponges of Canada and Alaska. Ottawa Nat., 14:153-72.
------. 1905. A new recent marine sponge *(Esperella bellabellensis)* from the Pacific coast of Canada. Ottawa Nat., 19:14-15.
Laubenfels, M. W. de. 1928. A new genus and species of sponge from Puget Sound. Publ. Puget Sound Biol. Station, 5:361-63.
------. 1932. The marine and fresh-water sponges of California. Proc. U. S. Nat. Mus., 81 (4):1-140.
------. 1948. The order Keratosa of the phylum Porifera--a monographic study. Occas. Pap. Allan Hancock Found., no. 3. 217 pp.
------. 1961. Porifera of Friday Harbor and vicinity. Pacific Sci., 15:192-202.
Okada, Y. 1932. Report on the hexactinellid sponges collected by the United States Fisheries Steamer "Albatross" in the northwestern Pacific during the summer of 1906. Proc. U. S. Nat. Mus., 81 (12):1-118.
Simpson, T. L. 1966. A new species of clathriid sponge from the San Juan Archipelago. Postilla, 103:1-7.

PHYLUM CNIDARIA

CLASS HYDROZOA

ORDER HYDROIDA

SUBORDER ANTHOMEDUSAE (=ATHECATA)

The hydroid stage of members of this group is generally athecate; that is, the perisarc stops at the base of the hydranth, instead of becoming expanded into a cuplike structure around the hydranth. (The only exceptions to this rule in our area are the genera *Garveia* and *Bimeria*, in which the hydranths are enclosed, and *Perigonimus* and *Bougainvillia*, in which they are partially enclosed.)

In the key to hydroids, couplets 3 to 13 cover the athecate genera likely to be found in this region. As far as most genera are concerned, it seems premature to attempt to prepare a key to species, as our fauna of hydroids is still rather poorly understood in spite of the extensive study of Fraser (1937). Those who wish to attempt to identify species will have to consult Fraser's keys and illustrations. The work of Naumov (1960) is admirable in many respects, and especially because his system of classification takes into consideration simultaneously the hydroid and medusa stages, when both exist in the same life history. Many of the genera and species dealt with by Naumov occur in our area.

Medusae of this group, if released to mature as free-swimming jellyfishes, are characterized by tall bells, gonads borne on the manubrium (instead of on the radial canals), and absence of marginal lithocysts. The genera and species reported from our area are included, along with Leptomedusae and members of the Trachylina, in the key to hydromedusae.

SUBORDER LEPTOMEDUSAE (=THECATA)

The hydroid stage of Leptomedusae is characterized by the presence of a hydrotheca, a cuplike expansion of the perisarc around the hydranth. The medusa stage, unless retained in the gonophore, is generally characterized by a shallow bell, gonads borne on the radial canals, and frequently also by marginal lithocysts. The keys to hydroids (couplets 14 to 35) and hydromedusae cover the respective stages of our representatives of this group.

SUBORDER LIMNOMEDUSAE

The medusa stage of *Gonionemus vertens* A. Agassiz, 1862 is abundant in summer. *Eperetmus typus* Bigelow, 1915 is rare. (See key to hydromedusae, choices 15a, 16a.)

Colonies of the two-tentacled hydroid stage of *Proboscidactyla flavicirrata* Brandt, 1838 are regularly found at the lips of the tubes of the sabellid polychaetes *Schizobranchia insignis* and *Pseudopotamilla occelata*. (See key to hydroids, choice 1a.) The small medusa stage is sometimes common. (See key to hydromedusae, choice 11a.)

ORDER TRACHYLINA

SUBORDER TRACHYMEDUSAE

Aglantha digitale (O. F. Müller, 1776) is the only species regularly found in our area. (See key to hydromedusae, choice 14a.)

SUBORDER NARCOMEDUSAE

Solmissus marshalli Agassiz & Mayer, 1902 and *Aegina citrea* Eschscholtz, 1829 occur in our area. (See key to hydromedusae, couplet 35.)

ORDER HYDROCORALLINA

SUBORDER STYLASTERINA

In some recent systems of classification of Hydrozoa, the stylasterines and milleporines (the latter not represented in our area) are placed in the suborder Anthomedusae of the order Hydroida. Naumov (1960) treats the Hydrocorallina (Hydrocorallia) as a suborder equivalent to the Anthomedusae.

Stylantheca porphyra Fisher, 1931, which forms pink, purple, or lavender encrustations on rocks, is abundant subtidally in our area; it is occasionally collected intertidally. Other species occur in deep water. Fisher (1931, 1938) and Naumov (1960) are the most useful references pertaining to stylasterines of this general region.

ORDER CHONDROPHORA

The members of this group are also sometimes placed in the Anthomedusae, because of histological similarities with *Tubularia* and its allies. In older classifications, they are assigned to the Siphonophora.

Velella velella (Linnaeus, 1758), the "purple sailor" or "by-thewind sailor," is occasionally carried by currents into our area, and then may be washed up on shore. However, it is rarely found except on the open coast or in bays that front on the open coast.

Key to Hydroids

1a Hydranths with two tentacles; found only near the openings of the tubes of certain sabellid polychaetes *(Schizobranchia, Pseudopotamilla) Proboscidactyla flavicirrata* Brandt, 1838 (Suborder Limnomedusae)
1b Hydranths with several to many tentacles 2
2a Hydranth not enclosed by a hydrotheca or pseudohydrotheca 3
2b Hydranth at least partially enclosed by a hydrotheca or pseudohydrotheca 10

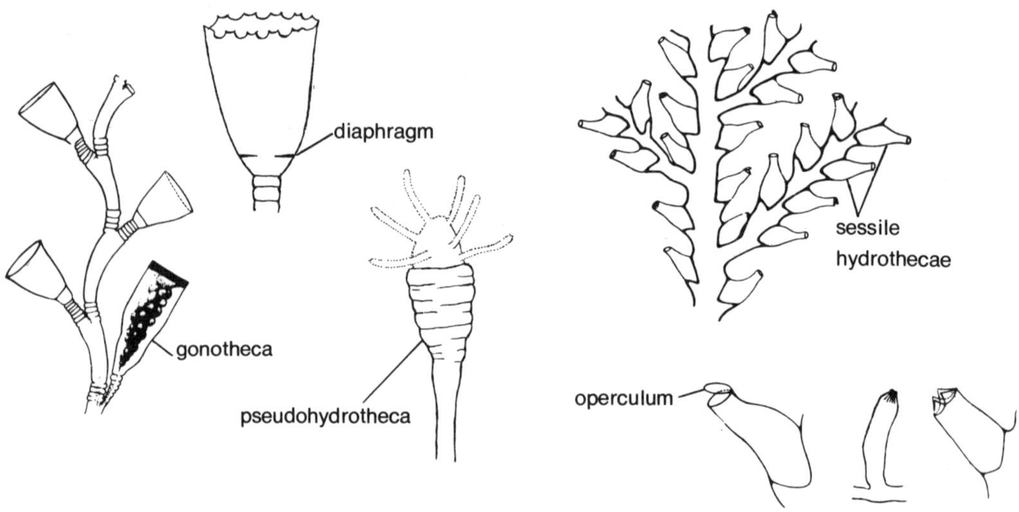

3a Tentacles on the hydranths terminating in knobs *Coryne*
3b Tentacles on the hydranths not terminating in knobs 4
4a Tentacles on the hydranths scattered over the surface, not in distinct circles 5
4b Tentacles on the hydranths arranged in 1 or 2 distinct circles 7
5a Stalks of hydranths generally unbranched; gonophores arising on the lower portions of the hydranths; polyps generally under 1 cm tall; strictly marine 6
5b Stalks of hydranths extensively branched; gonophores on the whole, arising well below the hydranths; polyps up to several centimeters tall; in brackish or nearly fresh water *Cordylophora caspia* (Pallas, 1771)
6a Gonophores numerous, in dense clusters, not liberating medusae; free-living *Clava multicornis* (Forskal, 1775)
6b Gonophores few (usually 1 to 3), liberating small medusae with 4 tentacles; found in the pharyngeal cavity of certain ascidians, especially *Ascidia paratropa* *Endocrypta huntsmani* (Fraser, 1911)
7a Tentacles of hydranths arranged in 2 circles 8
7b Tentacles of hydranths arranged in 1 circle 9
8a Polyps single; perisarc very thin; not likely to be found intertidally or on floats *Corymorpha*
8b Polyps forming clusters, joined at the base, and sometimes branching;

perisarc thick, often kinked; may be found intertidally and on floats, as well as subtidally *Tubularia*

9a Stalks of hydranths not enclosed by a perisarc (however, the stolons that encrust the substrate [usually a snail shell occupied by a hermit crab] have a perisarc, and this is generally spiny) *Hydractinia*

9b Stalks of hydranths enclosed by a perisarc, this characteristically strongly ringed *Eudendrium*

10a Perisarc enclosing or partially enclosing the hydranth constituting a pseudohydrotheca rather than a true hydrotheca (pseudohydrothecae are generally of inconsistent and somewhat irregular form, and show transverse wrinkles) 11

10b Perisarc enclosing or partially enclosing the hydranth constituting a true hydrotheca (of consistent form, and often with toothed margins, devices for closing the aperture, and other elaborations) 14

11a Pseudohydrotheca enclosing the lower portion of the hydranth flaring to some extent, but not forming a distinct cup 12

11b Pseudohydrotheca enclosing the lower portion of the hydranth elaborated into a distinct cup 13

12a Colony low (the hydranths generally less than 5 mm tall), consisting to a large extent of creeping stolons; stalks of hydranths branching only occasionally or not at all *Perigonimus*

12b Colony up to several centimeters tall; stalks of hydranths generally branched *Bougainvillia*

13a Perisarc around the gonophores disappearing before these are fully developed; hydranths usually bright orange *Garveia*

13b Perisarc around the gonophores persistent; hydranths not orange *Bimeria*

14a Hydrotheca small, not large enough to accommodate the hydranth when it contracts *Halecium*

14b Hydrotheca well developed, large enough to accommodate the hydranth when it contracts 15

15a Hydranths generally distinctly stalked (except in certain species of *Lafoea)* 25

15b Hydranths not stalked, but applied tightly to the stem or branches 16

16a Hydrotheca without an operculum; with nematophores (specialized small polyps in which nematocysts are concentrated; there is often a nearly constant number of these in relation to each hydranth) 23

16b Hydrotheca with an operculum consisting of 1 to 4 flaps; without nematophores 17

17a Hydranths arranged in a single series on one side of the branches; operculum consisting of a single flap *Hydrallmania*

17b Hydranths arranged in 2 series, on opposite sides of the branches, or in more than 2 series; operculum consisting of 1 or 2 flaps 18

18a Hydranths in 2 series, those on opposite sides of the branches arising at the same level and thus forming pairs 19

18b Hydranths in 2 or more series, alternating at least to some extent, not regularly forming pairs 20

19a Margin of the hydrotheca with 2 sharp, toothlike lobes; operculum consisting of 2 flaps *Dynamena*

19b Margin of the hydrotheca without toothlike lobes; operculum consisting of a single flap *Diphasia*

20a Operculum consisting of a single flap 21

20b Operculum consisting of 2 to 4 flaps 22

21a Operculum attached on the side of the hydrotheca that is nearer the axis of the branch *Abietinaria*

21b Operculum attached on the side of the hydrotheca that is farther from the axis of the branch *Thuiaria* (including *Selaginopsis*)

22a Margin of the hydrotheca with 2 toothlike lobes; operculum consisting of 2 flaps *Sertularia*

22b Margin of hydrotheca with 3 or 4 toothlike lobes; operculum consisting of 3 or 4 flaps *Sertularella*

23a Thecae of nematophores articulating with the branch from which they arise and thus movable to some extent *Plumularia*

23b Thecae of nematophores not articulating with the branch from which they arise, and thus not movable 24

24a Each hydranth-bearing unit of a branch provided with 2 nematophores; gonophores situated within corbulae, which consist of several pairs of leaflike expansions located on some of the branches *Aglaophenia*

24b Each hydranth-bearing unit of a branch provided with 3 nematophores; gonophores not situated within corbulae, but associated with slender branches (phylactogonia) that lack hydranths but have nematophores *Cladocarpus*

25a Hydrotheca with an operculum of 4 or more converging flaps 26

25b Hydrotheca without an operculum 28

26a Opercular flaps arranged in such a way that they form 2 sloping surfaces *Stegopoma*

26b Opercular flaps arranged in such a way that they form a conical or pyramidal configuration, or a funnel-shaped concavity 27

27a Basal portions of opercular flaps sharply demarcated from the margin of the hydrotheca *Calicella*

27b Opercular flaps merely extensions of the margin of the hydrotheca *Campanulina*

28a Hydrotheca bell-shaped or goblet-shaped; a diaphragm always present in the lower part of the hydrotheca 29

28b Hydrotheca generally tubular, but sometimes bell-shaped or goblet-shaped; diaphragm absent (except in species of *Lictorella*, which have a delicate diaphragm, but which are unlike most diaphragm-bearing hydroids in having the hydrotheca more convex on one side than elsewhere 31

29a Colony consisting largely of stolons applied to the substrate, the free branches rarely bearing more than 2 or 3 hydranths *Campanularia* (including *Clytia* and *Eucopella*)

29b Colony consisting largely of free stems (these may be fascicled) that bear many hydranths 30

30a Colony branching sympodially, the hydranths being produced alternately on one side of a branch and then on the other; stems not fascicled *Obelia* (including *Gonothyraea*, which, instead of releasing free-medusae, retains them in a cluster at the tip of the gonophore)

30b Colony branching irregularly; stems fascicled *Verticillina*

31a Stalks of hydranths distinct, smooth or annulated; hydrothecae not becoming partially fused to the stems that produce them *Bonneviella*

31b Stalks of hydranths spirally twisted or lacking; hydrothecae often becoming partially fused to the stems that produce them 32

32a Hydrotheca with a delicate diaphragm *Lictorella*

32b Hydrotheca without a diaphragm 33

33a Hydrotheca like a slender goblet, either on a stalk or directly attached to the stolon from which it arises; wall of the hydrotheca not adherent to the stolon *Lafoea*

33b Hydrotheca decidedly tubular, with a short stalk or none; hydrotheca either adherent to the stolon or actually buried partially or almost completely in the mass of stolons 34

34a Colony creeping (usually on other hydroids); lower half of the hydrotheca adherent to the stolon *Filellum*

34b Colony branching free of the substrate; hydrotheca either partially adherent to the stolon or embedded in it partially or almost completely 35

35a Hydrothecae in 2 or 4 somewhat definite rows with respect to the stem from which they arise; hydrotheca bent back against the stem, so that a portion of the wall above the base is adherent to the stem (however, the lowermost and uppermost portions of the hydrotheca are free of the stem) *Cryptolaria*

35b Hydrothecae not in obvious rows, or else in more than 4 rows; hydrotheca buried almost completely in the stem *Grammaria*

Key to Hydromedusae

1a Tentacles arising at the margin of the bell 2

1b Tentacles arising decidedly above the margin of the bell 35

2a With not more than 4 tentacles 3

2b With more than 4 tentacles 10

3a With 2, 3, or 4 tentacles arising individually from widely separated tentacular bulbs; maximum height of bell usually exceeding 6 mm 4

3b With up to 3 (rarely 4) tentacles arising from one simple or compound tentacular bulb (medusae are produced asexually from this bulb also; three other bulbs around the margin of the bell are rudimentary); maximum height of bell about 4 mm *Tubularia prolifer* (L. Agassiz, 1862) *(Hybocodon prolifer)*

4a With 2 tentacles situated on opposite sides of the bell *Stomotoca atra* L. Agassiz, 1862

4b With 3 or 4 tentacles 5

5a With 3 tentacles, 1 of which is longer than the other 2, and a bulb from which no tentacle arises *Corymorpha tentaculata* (Linko, 1904) *(Euphysa tentaculata)*

5b With 4 equally developed tentacles 6

6a With a pigmented eyespot in the bulb of each tentacle 7

6b Without a pigmented eyespot in the bulb of each tentacle *Corymorpha flammea* (Linko, 1904) *(Euphysa flammea)*

7a Each tentacle tipped with a conspicuous knob in which nematocysts are concentrated, and with 3 to 6 rings of nematocysts just proximal to the knob *Dipurena halterata* (Forbes, 1846)

7b Tentacles not terminating in knobs, and without distinct rings of nematocysts just proximal to the tip 8

8a Manubrium capable of being extended beyond the margin of the bell for a distance equal to at least the height of the bell 9

8b Manubrium not capable of being extended much beyond the margin of the bell *Coryne rosaria* L. Agassiz, 1862 *(Sarsia rosaria)*

9a Aboral end of the bell simply rounded, without a distinct apical prolongation; very common *Coryne tubulosa* (M. Sars, 1835) *(Sarsia tubulosa)*

9b Aboral end of the bell with a distinct apical prolongation, so that the bell is more or less conical; rare *Coryne princeps* (Haeckel, 1879) *(Sarsia princeps)*

10a Tentacles (usually 9) branched, one branch of each ending in a sucker, some of the others in batteries of nematocysts *Cladonema californicum* Hyman, 1947

10b Tentacles not branched 11

11a Radial canals (there are usually 4 to 6 of these, rarely 7, originating from the stomach) branching more or less dichotomously, the ultimate branches reaching the margin of the bell *Proboscidactyla flavicirrata* (Brandt, 1838)

11b Radial canals not branching dichotomously (however, they may branch into numerous lateral diverticula) 12

12a With 4 or 8 radial canals 13

12b With numerous radial canals *Aequorea aequorea* (Forskål, 1775)

13a With 4 radial canals 15

13b With 8 radial canals 14

14a Bell about twice as tall as wide; gonads sausage-shaped, much shorter than the radial canals *Aglantha digitale* (O. F. Müller, 1776)

14b Bell about as tall as wide; gonads sinuous, almost the full length of the radial canals *Melicertum octocostatum* (M. Sars, 1835)

15a Each tentacle tipped with a distinct knob in which nematocysts are concentrated *Eperetmus typus* Bigelow, 1915

15b Tentacles not tipped with knobs 16

16a Tentacles with adhesive pads some distance from their tips, and characteristically angled where the pads are located *Gonionemus vertens* A. Agassiz, 1862

16b Tentacles without adhesive pads some distance from their tips, and not characteristically angled 17

17a Radial canals with conspicuous lateral diverticula whose length is at least twice the diameter of the canals 18

17b Radial canals without conspicuous lateral diverticula (if the margins of the canals are irregular, they do not form distinct diverticula whose length appreciably exceeds the diameter of the canals) 20

18a Bell distincly taller than wide; gonads fingerlike, in 4 groups, hanging down into the cavity of the bell *Polyorchis penicillatus* (Eschscholtz, 1829)

18b Bell wider than tall; gonads closely associated with the radial canals and their diverticula 19

19a With up to about 500 tentacles, these without eyespots at their bases; diameter of bell up to about 9 cm *Ptychogena lactea* A. Agassiz, 1865

19b With up to about 4000 tentacles, these with eyespots at their bases; diameter regularly exceeding 10 cm, and sometimes exceeding 20 cm *Staurophora mertensii* Brandt, 1835

20a With only 8 substantial tentacles (4 of these may be shorter than the others, however) and a number of rudimentary tentacles 21

20b With more than 8 substantial tentacles 22

21a With 8 substantial tentacles of approximately equal length, and about 40 to 60 rudimentary tentacles; bell wider than tall, without an apical projection *Tiaropsidium kelseyi* Torrey, 1909

21b With 4 longer tentacles in line with the radial canals and 4 shorter tentacles alternating with these, plus a few rudimentary tentacles; bell taller than wide, with a conical apical projection *Halitholus pauper* Hartlaub, 1913

22a Tentacles in 4 or 8 clusters 23

22b Tentacles not in clusters 24

23a Tentacles in 4 clusters, with from 15 to 40 in each cluster *Bougainvillia* sp.

23b Tentacles in 8 clusters, with clusters of 3 and 5 tentacles alternating *Rathkea octopunctata* (M. Sars, 1835)

24a Tentacles of 2 lengths, each longer one alternating with 1 or more shorter ones 25

24b Tentacles not of 2 lengths (however, there may be variation in the length of the tentacles, and rudimentary marginal bulbs may also be present) 26

25a With up to about 50 longer tentacles, each of these alternating with a series of several shorter ones (cirri) *Mitrocomella sinuosa* (Foerster, 1923)

25b With about 180 to 240 longer tentacles, each of these alternating with 1 short tentacle (cirrus) *Mitrocoma discoidea* Torrey, 1909

26a Bell about as tall as, or taller than, wide; upper portion of bell generally either distinctly set off by a constriction, or somewhat prolonged (in *Halimedusa typus*, it is merely thickened) 27

26b Bell decidedly wider than tall; upper portion of bell not distinctly set off by a constriction or prolonged 31

27a Upper portion of bell distinctly set off by a constriction; includes 1 common species (*Leuckartiara octona*) 29

27b Upper portion of bell not set off by a constriction, but prolonged as a bluntly conical elevation or simply somewhat thickened; species rare in our area 28

28a Rim of mouth with knobs of nematocysts; upper portion of bell not conspicuously prolonged, but simply thickened; with about 45 tentacles *Halimedusa typus* Bigelow, 1916

28b Rim of mouth without knobs of nematocysts; upper portion of bell prolonged as a bluntly conical elevation; with up to 140 (usually 60 to 100) tentacles *Leuckartiara breviconis* (Murbach & Shearer, 1902)

29a With rudimentary marginal bulbs alternating with well-developed (but not necessarily equal) tentacles 30

29b Without rudimentary marginal bulbs *Leuckartiara nobilis* Hartlaub, 1913

30a With 12 to 24 tentacles (usually about 16); rather common *Leuckartiara octona* Fleming, 1823

30b With 30 to 40 tentacles; uncommon *Catablema vesicarium* (A. Agassiz, 1862)

31a Gonads ovoid or nearly spherical, occupying only a short section of each radial canal *Obelia* spp.

31b Gonads elongated, associated with the radial canals for a substantial part of their length 32

32a With fewer than 100 tentacles 33

32b With more than 100 tentacles 34

33a With about 32 tentacles; gonads generally more than half the length of the radial canals *Phialidium hemisphericum* (Linnaeus, 1767)

33b With about 60 tentacles; gonads occupying about half the length of the radial canals *Phialidium gregarium* (L. Agassiz, 1862)

34a Mouth on a long stalk into which the radial canals descend; with about 150 to 230 tentacles; diameter of bell not exceeding 1.5 cm *Eutonina indicans* (Romanes, 1876)

34b Mouth not on a long stalk; with more than 250 tentacles; diameter of bell regularly between 2 and 5 cm, sometimes longer *Halistaura cellularia* (A. Agassiz, 1865)

35a Bell up to about 6 cm in diameter; with up to about 20 tentacles *Solmissus marshalli* Agassiz & Mayer, 1902

35b Bell not over 2.5 cm in diameter; with 4 tentacles *Aegina citrea* Eschscholtz, 1829

ORDER SIPHONOPHORA

George O. Mackie
University of Victoria, British Columbia

1a With a gas-filled float, and numerous nectophores arranged in two rows behind this; nectophores, stem, and gastrozooids with distinct chromatophores (orange when expanded, plum-colored when contracted); terminal stem groups not becoming detached and living apart from the colony (Suborder Physonectae) *Nanomia cara* Agassiz, 1865

1b Without a gas-filled float, and with only 1 or 2 nectophores; without distinct chromatophores (although pigment may be present); terminal stem groups becoming detached and living apart from the colony as "eudoxids" (Suborder Calycophorae) 2

2a With 2 soft, gelatinous, rounded nectophores, each up to 5 cm long, arranged "back to back;" bracts of eudoxids with 4 major canals running through the jelly *Praya reticulata* Bigelow, 1911 (Only short portions of the stem of this large species are likely to be encountered. The tentacular nematocyst batteries and bases of the gastrozooids are distinctive in being cadmium yellow.)

2b With 1 or 2 firm, angled or ridged nectophores not over 2.5 cm long (when there are 2, they are "in line," rather than "back to back"); bracts of eudoxids without canals 3

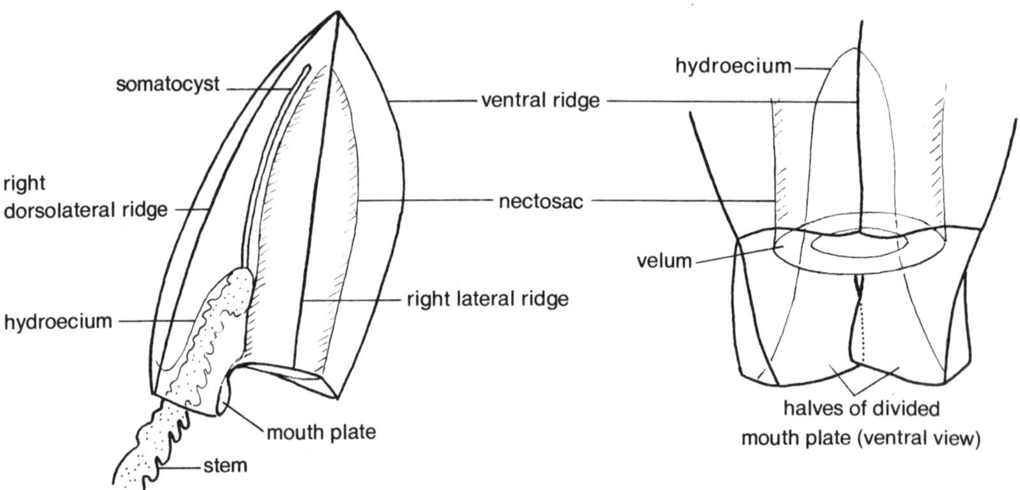

Muggiaea

3a Nectophore(s) with 5 prominent longitudinal ridges; anterior nectophore (if 2 are present) sharply pointed and about one and a half times as long as the posterior one; mouthplate of nectophore(s) divided 4

3b Nectophore(s) without prominent ridges; anterior nectophore rounded and about twice as long as the posterior one; mouthplate of nectophore(s) not divided *Dimophyes arctica* (Chun, 1897)

4a With a single nectophore up to about 5 mm long; somatocyst long, reaching approximately to the level of the apex of the nectosac *Muggiaea atlantica* Cunningham, 1892

4b Usually with 2 nectophores, the anterior one up to 2.5 cm long; somatocyst ending some distance below the apex of the nectosac *Lensia conoidea* (Keferstein and Ehlers, 1860)

Bigelow, H. B. 1913. Medusae and Siphonophorae collected by the U. S. Fisheries steamer "Albatross" in the northwestern Pacific, 1906. Proc. U. S. Nat. Mus., 44:1-119.

------. 1915. *Eperetmus*, a new genus of Trachymedusae. Proc. U. S. Nat. Mus., 49:399-404.

------. 1916. *Halimedusa*, a new genus of Anthomedusae. Trans. Royal Soc. Canada, Sect. 4, 10:91-6.

Fields, W. G., & G. O. Mackie. 1971. Evolution of the Chondrophora: evidence from behavioral studies on *Velella*. J. Fish. Res. Bd. Canada, 28:1595-1602.

Fisher, W. K. 1931. Californian hydrocorals. Ann. Mag. Nat. Hist., ser. 10, 8:391-99.

------. 1938. Hydrocorals of the north Pacific Ocean. Proc. U. S. Nat. Mus., 84:493-554.

Foerster, R. E. 1923. The hydromedusae of the west coast of North America, with special reference to those of the Vancouver Island region. Contrib. Canad. Biol., n.s., 1:219-77.

Fraser, C. McL. 1937. Hydroids of the Pacific coast of Canada and the United States. Toronto: University of Toronto Press. 207 pp.

Kramp, P. L. 1961. Synopsis of the medusae of the world. J. Mar. Biol. Assoc. United Kingdom, 40:1-469.

Mackie, G. O. 1964. Analysis of locomotion in a siphonophore colony. Proc. Royal Soc. London, B, 159:366-91.

Mackie, G. O., & G. V. Mackie. 1963. Systematic and biological notes on living hydromedusae from Puget Sound. Nat. Mus. Canada Bull. 199:63-84.

Mayer, A. G. 1910. Medusae of the World. Vols. 1, 2: Hydromedusae. Carnegie Inst. Washington Publ. 109:1-498, i-xv.

Naumov, D. V. 1960. [Hydroids and hydromedusae of marine, brackish and freshwater basins of the U.S.S.R.] Opredeliteli po Faune SSSR, 70. 626 pp. (in Russian). (English translation, 1969. Jerusalem: Israel Program for Scientific Translation.)

Russell, F. S. 1953. The Medusae of the British Isles. Anthomedusae, Leptomedusae, Limnodedusae, Trachymedusae and Narcomedusae. Cambridge: Cambridge University Press. xiii + 530 pp.

Stepan'iants, S. D. 1967. [Siphonophora of seas of the USSR and of the northern part of the Pacific Ocean.] Opredeliteli po Faune SSSR, 96. 216 pp. (in Russian).

Totton, A. K. 1965. A synopsis of the Siphonophora. London: British Museum (Natural History). vii + 230 pp.

CLASS SCYPHOZOA

ORDER DISCOMEDUSAE

SUBORDER SEMAEOSTOMAE

1a Tentacles short, forming a fringe at the margin of the bell; with 8 rhopalia, these with a small tentaclelike lappet on both sides 2

1b Tentacles long, arising singly from the clefts between certain lobes or in groups of several to many from the underside of the lobes, or both; with 8 or 16 rhopalia, these without tentaclelike lappets 3

2a Margin of the bell divided into 8 equal lobes *Aurelia aurita* (Linnaeus, 1758)

2b Margin of the bell with 16 lobes (there are 8 primary lobes, but each of these is notched) *Aurelia labiata* Chamisso & Eysenhardt, 1821

3a All tentacles arising from the underside of the bell 4

3b At least some of the tentacles arising in the clefts between the marginal lobes 5

4a With 16 rounded lobes alternating with smaller lobes that are divided in such a way that they resemble fishtails (the 16 rhopalia are situated on these); with up to about 25 tentacles arranged in a single row on the underside of the rounded lobes *Phacellophora camtschatica* Brandt, 1838

4b With a total of 16 more or less equal marginal lobes, arranged in 8 pairs (the 8 rhopalia are situated between the pairs); with up to about 150 tentacles arranged in up to 4 rows having a U-shaped pattern and extending from one lappet to another around each cleft where a rhopalium is located *Cyanea capillata* (Linnaeus, 1758)

5a Bell without radial lines or bands; marginal lobes with reddish spots *Chrysaora helvola* Brandt, 1838

5b Bell with dark radial lines or bands; marginal lobes without spots 6

6a Exumbrellar surface with a dark ring from which broad dark bands radiate *Chrysaora melanaster* Brandt, 1838

6b Exumbrellar surface without a dark ring, and usually without radiating dark bands *Chrysaora quinquecirrha* (Desor, 1851)

ORDER STAUROMEDUSAE

These are attached, by an aboral stalk, to *Zostera* and various brown algae. We have only a few species, but the identity of some of these is uncertain. Naumov (1961) has a helpful account of this group.

1a At least the outer tentacles with cushionlike swellings at the bases *Thaumatoscyphus* sp.

1b None of the tentacles with cushionlike swellings at their bases *Haliclystus auricula* (Rathke, 1806)

There may be another species of *Haliclystus* in our area.

Mayer, A. G. 1910. Medusae of the World, vol. III. Carnegie Institution of Washington Publ. 109:i-iv, 499-735.

Naumov, D. V. 1961. [Scyphozoan medusae of seas of the U.S.S.R.] Opredeliteli po Faune SSSR, 75. 98 pp. (in Russian).

Russell, F. S. 1970. Medusae of the British Isles. II. Pelagic Scyphozoa with a Supplement to the First Volume on Hydromedusae. Cambridge: Cambridge University Press. xii + 284 pp.

CLASS ANTHOZOA

Arthur E. Siebert, Jr.
University of California, Berkeley

1a Polyps with 8 pinnately branched tentacles — Subclass Octocorallia
1b Polyps not with 8 tentacles, and tentacles not pinnately branched — 2
2a With more than 12 tentacles arranged in 2 distinct cycles only; inhabiting tubes consisting of nematocysts embedded in a matrix of slippery mucus — Subclass Ceriantipatharia, Order Ceriantharia
2b With not more than 12 tentacles in the first 2 cycles, and usually with additional cycles of tentacles; not inhabiting tubes consisting of nematocysts embedded in a mass of slippery mucus — Subclass Zoantharia

SUBCLASS OCTOCORALLIA

1a Colony encrusting, the polyps arising singly from creeping stolons — Order Stolonifera
1b Colony in the form of a rounded or lobed fleshy mass, or featherlike or irregularly branched — 2
2a Colony a rounded or lobed fleshy mass in which the polyps are embedded — Order Alcyonaria
2b Colony featherlike or irregularly branched — 3
3a Colony featherlike, the lower portion swollen and anchoring the colony in mud or sand; rachis of the colony supported by a hard internal rod — Order Pennatulacea
3b Colony irregularly branched, attached to a hard substrate; branches of the colony supported by an internal core of horny material — Order Gorgonacea

ORDER STOLONIFERA

Clavularia sp. (largely subtidal)

ORDER ALCYONARIA

Gersemia rubiformis (Pallas) (subtidal)

ORDER GORGONACEA

Paragorgia arborea (Linnaeus, 1758) (at depths of 100 m and greater) Other species will be taken by dredging on the continental shelf, at depths of about 300 m and greater.

ORDER PENNATULACEA

1a Rachis and branches thick and bushy, pale to vivid orange — *Ptilosarcus gurneyi* (Gray, 1860)
1b Rachis and branches slender, not especially bushy, white or almost white — 2
2a Rachis smooth to the touch, without projecting calcareous spicules — *Virgularia* sp.

2b Rachis rough to the touch, owing to projecting calcareous spicules *Stylatula elongata* (Gabb, 1863)

SUBCLASS CERIANTIPATHARIA

ORDER CERIANTHARIA

Pachycerianthus fimbriatus (McMurrich, 1910) (subtidal)

SUBCLASS ZOANTHARIA

1a Colonial, the bases of the polyps in each group connected; column with sand or other foreign material embedded in it Order Zoanthiniaria

1b Not colonial (although some species form large aggregations); column without sand or other foreign material actually embedded in it, although such material may adhere to the outside of it, and an external calcareous skeleton may be present 2

2a Polyps with a calcareous external skeleton Order Scleractinia

2b Polyps without a calcareous skeleton Order Actiniaria

ORDER ZOANTHINIARIA

Epizoanthus scotinus Wood, 1958 (intertidal and subtidal)

ORDER SCLERACTINIA

1a Polyps bright orange; inner edges of septa not linked to vertical pillars; intertidal as well as subtidal *Balanophyllia elegans* Verrill, 1864

1b Polyps beige, brown, or pink; inner edges of certain septa linked to vertical pillars; strictly subtidal, and generally found only in deep water *Caryophyllia alaskensis* Vaughan, 1941

ORDER ACTINIARIA

1a Somewhat wormlike, burrowing; base rounded, never attached to a hard substrate 2

1b Not wormlike, and seldom burrowing; base flat, attached to a hard substrate 5

2a With 10 tentacles *Halcampa decemtentaculata* Hand, 1954

2b With 12 tentacles 3

3a Column tan to brown; with folds (conchula) on the oral disk around the siphonoglyph *Peachia quinquecapitata* McMurrich, 1913

3b Column translucent and colorless to opaque and beige; without folds on the oral disk around the siphonoglyph 4

4a Lips and pharynx beige, white, or occasionally brown, never red; base rounded when contracted *Halcampa* sp.

4b Lips and pharynx usually red; base flattened when contracted *Halcampoides purpurea* (Studer, 1878)

5a With threadlike structures (acontia) that can be discharged through pores (cinclides) or breaks in the body wall when the anemone is disturbed or damaged 6

5b Without threadlike structures that can be discharged through pores or breaks in the body wall 9

6a Margin of oral disk lobed; tentacles slender, in large specimens numbering 200 or more; height regularly exceeding 5 cm; column white, orange, or brown, without contrasting stripes
Metridium senile (Linnaeus, 1767)

6b Margin of disk not lobed; tentacles relatively thick, usually fewer than 100; height not exceeding 5 cm; column green to brown, with orange stripes, or white, orange, yellow, or gray, and without stripes 7

7a Column green to brown, with about 7 to 19 orange stripes
Haliplanella luciae (Verrill, 1898)

7b Column without orange stripes, and not green, but may be white, orange, yellow, gray, or other colors 8

8a Column beige or cream; tentacles beige or cream, with brown bands or spots; one directive tentacle may be solid white
unidentified or undescribed sagartiid

8b Color variable; orange, yellowish, reddish, gray, cream, or brown; tentacles all of one color
unidentified *Diadumene* spp. and juvenile *Metridium senile*

9a Column with tubercles, these often accumulating sand, gravel, and bits of shell 10

9b Column without tubercles, and usually without adherent material 17

10a Margin of oral disk with white or yellow spherules, these often hidden by the tentacles 11

10b Margin of oral disk without spherules 13

11a Tubercles restricted to upper portion of the column; tentacles usually long and slender, with white bands on a gray, brown, blackish, or green background; buried in sand or shell gravel, with the base attached to a solid substrate such as a rock or shell; without symbiotic algal cells in the tissues
Anthopleura artemisia (Pickering, in Dana, 1848)

11b Entire column covered with tubercles; usually with symbiotic algal cells (green or olive) in the tissues 12

12a Column green to white; tubercles usually in distinct longitudinal rows; tentacles with pink tips; height up to about 5 cm; often in aggregating masses, and frequently buried by sand covering rocks to which they are attached *Anthopleura elegantissima* (Brandt, 1835)

12b Column green or olive green; tubercles usually not in distinct longitudinal rows; tentacles uniform in color and not pink-tipped; height regularly exceeding 5 cm; solitary, and not often buried in sand
Anthopleura xanthogrammica (Brandt, 1835)

13a Column white, pink, or yellow; tentacles with fine, irregular lines; oral disk pink, yellow, or white, with pink or yellow lines radiating from the mouth to the margin; strictly subtidal *Cribrinopsis* sp.

13b Column not white, pink, or yellow; tentacles without fine lines, though they may have coarse stripes or bands of pigment; lines on oral disk, if present, not pink or yellow; subtidal or intertidal 14

14a Column green or pink, with white tubercles; oral disk with conspicuous white stripes radiating from the mouth to the margin
Cnidopus ritteri (Torrey, 1902)

14b Column greenish, olive, brown, or red; oral disk without radiating stripes 15

15a Column covered with tubercles that accumulate sand, gravel, and bits of shell, usually partially buried in coarse sand or gravel; column red; oral disk greenish *Tealia coriacea* (Cuvier, 1798)

15b Tubercles not tending to accumulate sand or other foreign material; column brown or red, or greenish or olive with red blotches or streaks; oral disk not greenish 16

16a Column red; column with a few white tubercles arranged in longitudinal rows; mostly subtidal *Tealia lofotensis* (Danielssen, 1890)

16b Column uniformly brown or red, or greenish or olive with red streaks and blotches; column without white tubercles or tubercles arranged in rows; intertidal, subtidal, and on floats
Tealia crassicornis (O. F. Müller, 1776)

17a Column orange, white, or white with orange blotches; tentacles white or orange, sometimes with orange bands 18

17b Column not white or orange, or white with orange blotches; tentacles without orange bands 19

18a Tentacles with white spots at their bases; column as wide as tall; usually attached to shells of *Modiolus*
Stomphia coccinea (O. F. Müller, 1776)

18b Tentacles without white spots at their bases; column taller than wide; usually not on shells of *Modiolus* *Stomphia didemon* Siebert, 1973

19a Oral disk and base of column with radiating white lines (color of column brown, green, or red); regularly with adherent young at the base of the column *Epiactis prolifera* Verrill, 1869

19b Oral disk and column without radiating white lines; without adherent young at the base of the column 20

20a Entire body dark green, with no conspicuous markings; in the upper intertidal region unidentified sp.

20b Column brown or red, or greenish or olive with red blotches or streaks; tentacles with transverse bands of various colors; in the lower intertidal region, subtidal, and on floats
Tealia crassicornis (O. F. Müller, 1776)

Arai, M. N. 1971. *Pachycerianthus* (Ceriantharia) from British Columbia and Washington. J. Fish. Res. Bd. Canada, 28:1677-1680.

Carlgren, O. 1936. Some west American sea anemones. J. Wash. Acad. Sci., 26:16-23.

------. 1949. A survey of the Ptychodactaria, Corallimorpharia and Actiniaria. Kungl. Svenska Vetenskap. Handl., ser. 4, 1:1-121.

Hand, C. 1955. The sea anemones of central California. Part I. The corallimorpharian and athenarian anemones. Wasmann J. Biol., 12 (1954):345-75.

------. 1955. The sea anemones of central California. Part II. The endomyarian and mesomyarian anemones. Wasmann J. Biol., 13:37-99.

------. 1956. The sea anemones of central California. Part III. The acontiarian anemones. Wasmann J. Biol., 13 (1955):189-251.

Hickson, S. V. 1915. Some Alcyonaria and a *Stylaster* from the west coast of North America. Proc. Zool. Soc. London, 1915:541-57.

Kükenthal, W. 1913. Über die Alcyonarienfauna Californiens und ihre tiergeographischen Beziehungen. Zool. Jahrb., Abt. Syst., 35:219-70.

Moore, R. C. (ed.). 1956. Treatise on Invertebrate Paleontology, Part F. Lawrence, Kansas: Geological Society of America and University of Kansas Press. 498 pp.

Siebert, A. E., Jr. 1973. A description of the sea anemone *Stomphia didemon* sp. nov. and its development. Pacific Sci., 27:363-76.

Stephenson, T. A. 1928, 1935. The British Sea Anemones, vols. I and II.

London: Ray Society. xiv + 148 p., xii + 426 pp.
Torrey, H. B. 1902. Papers from the Harriman Alaska Expedition, XXX. Anemones, with a discussion of variation in *Metridium*. Proc. Wash. Acad. Sci., 4:373-410.
Vaughan, T. W., and V. W. Wells. 1943. Revision of the suborders, families, and genera of the Scleractinia. Spec. Pap. Geol. Soc. Amer., 44:1-363.

PHYLUM CTENOPHORA

Although only a few species of ctenophores have been reported from this region, not even half of these can be identified with certainty. The main reason for our predicament is that specimens collected in our area have not been compared carefully with specimens known to belong to described species. Complicating matters further is the fact that some of the early descriptions and illustrations cannot be reconciled perfectly with actual specimens. In addition, some of our ctenophores seem not to have been described. The references provided will at least be helpful in confirming the identification of some more or less cosmopolitan forms, and in recognizing the genera to which ctenophores not in the key may belong.

1a Body resembling a sac, without tentacles and not divided into a pair of large lobes at the oral end; principal canals of the digestive tract branching irregularly, some of the branches anastomosing (Class Nuda) *Beroë* spp.

1b Body solid, not saclike, either with a pair of conspicuous tentacles arising within sheaths on opposite sides of the body, or with a pair of large lobes at the oral end (in which case there are small tentacles of 2 types in the cavity enclosed by the oral lobes) (Class Tentaculata) 2

2a Body divided into a pair of large lobes at the oral end; tentacles small, of 2 types (a pair of branched tentacles in sheaths near the mouth, and numerous small tentacles in tracts leading toward the mouth; both types are in the cavity enclosed by the oral lobes) *Bolinopsis ?microptera* (A. Agassiz, 1865)

2b Body not divided into a pair of large lobes at the oral end; tentacles conspicuous when extended, arising from sheaths on opposite sides of the body some distance from the mouth 3

3a Tentacles with side branches 4

3b Tentacles without side branches 6

4a Body elongate, approximately cucumber-shaped; length sometimes attaining 10 cm *Hormiphora cucumis* (Mertens, 1833)

4b Body spherical, ovoid, or pyriform; length not exceeding 2 cm 5

5a With red chromatophores concentrated along the bases of the comb rows; color of the animal as a whole often pinkish; uncommon *Pleurobrachia ?pileus* O. F. Müller, 1776

5b Essentially spherical; without red chromatophores concentrated along the bases of the comb rows; usually colorless, except for a pinkish tinge on the tentacles; common *Pleurobrachia bachei* A. Agassiz, 1860

6a Tentacles with nematocysts, but without colloblasts; tentacle sheaths with orange to brick-red pigment *Euchlora ?rubra* (Kölliker, 1853)

6b Tentacles without nematocysts, but with colloblasts; tentacle sheaths not noticeably pigmented (but there may be some orange-brown pigment near the mouth and gonads) *Dryodora ?glandiformis* (Mertens, 1833)

Agassiz, A. 1865. North American Acalephae. Mem. Mus. Comp. Zool. Harvard College, 1 (2):i-xiv, 1-234.
Chun, C. 1880. Die Ctenophoren des Golfes von Neapel und der angrenzenden Meeres-Abschnitte. Fauna und Flora des Golfes von Neapel, 1. xvi + 313 pp.
Komai, T. 1918. On ctenophores of the neighborhood of Misaki. Annot. Zool. Japon., 9:451-73.
Mayer, A. G. 1912. Ctenophores of the Atlantic coast of North America. Carnegie Inst. Washington Publ. 162. 115 pp.
Vanhoffen, E. Ctenophoren. Nordisches Plankton, Zool., 6, Abt. XI:1-7.

PHYLUM PLATYHELMINTHES

CLASS TURBELLARIA

SUBCLASS ARCHOOPHORA

ORDER ACOELIDA

Acoels are generally abundant in bottom sediments, as well as on algae and colonies of invertebrates that accumulate detritus or encourage growths of diatoms. They are found in intertidal as well as subtidal situations. The fauna of acoels in our region is extensive, but only 3 species have been described.

A key will not be offered, because of the risk that it might lead to incorrect identification of undescribed species. The monograph of Dörjes (1968) is very helpful for recognition of families and genera, and Kozloff (1965) illustrates the 3 species recorded from this area.

Family Otocelididae
Otocelis luteola (Kozloff, 1965)
Family Convolutidae
Raphidophallus actuosus Kozloff, 1965
Diatomovora amoena Kozloff, 1965

Dörjes, J. 1968. Die Acoela (Turbellaria) der deutschen Nordseeküste und ein neues System der Ordnung. Zeitschr. Zool. Syst. Evolutionsforsch., 6:56-452.
Kozloff, E. N. 1965. New species of acoel turbellarians from the Pacific coast. Biol. Bull., 129:151-66.
------. 1972. Selection of food, feeding, and physical aspects of digestion in the acoel turbellarian *Otocelis luteola*. Trans. Amer. Micr. Soc., 91:556-65.

ORDER POLYCLADIDA

John V. Holleman
Merritt College, Oakland, California

1a Ventral surface provided with a sucker (this is independent of the digestive system, and lies behind the female gonopore); tentacles, when present, situated along the margin of the body at the anterior end (Suborder Cotylea) ... 16

1b Ventral surface without a sucker; tentacles, when present, of the nuchal type, situated above the brain and some distance behind the anterior end (Suborder Acotylea) 2

2a A band of eyespots present along the entire margin of the body (eyespots also present on the tentacles, at the bases of the tentacles, and in the region of the brain); length up to about 10 cm
Kaburakia excelsa Bock, 1925

2b Eyespots restricted to the tentacles and the region of the brain; length generally less than 5 cm 3

3a With nuchal tentacles 4

3b Without nuchal tentacles 5

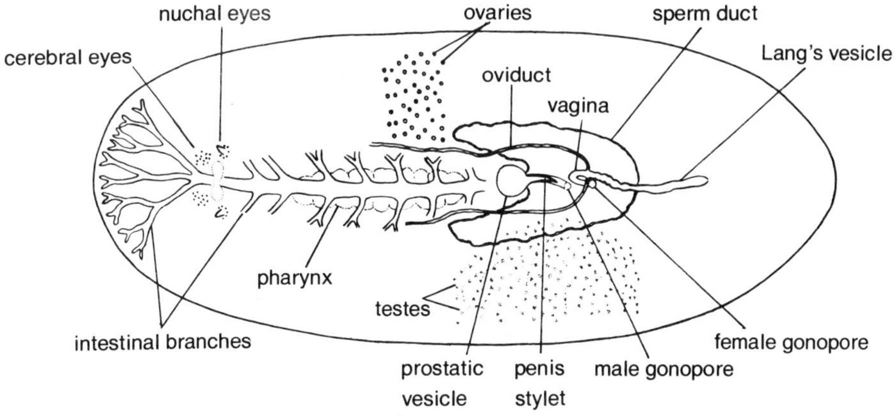

(dorsal view)

4a Chitinized penis stylet present and conspicuous; Lang's vesicle absent
Phyllaplana viridis (Freeman, 1933)

4b Penis stylet absent; Lang's vesicle (terminal dilation of a posteriorly directed diverticulum of the vagina) present
Pseudostylochus burchami (Heath & McGregor, 1912)

5a Chitinized penis stylet present 6

5b Penis stylet absent 9

6a Penis stylet straight or arched, never coiled 7

6b Penis stylet very long and coiled
Notoplana longastyletta (Freeman, 1933)

7a Penis stylet straight; gonopore single, serving both male and female reproductive systems *Notoplana inquieta* (Heath & McGregor, 1912)

7b Penis stylet arched; male and female gonopores separate 8

8a Penis stylet about as long as the prostatic vesicle (a glandular bulbous structure between the short penis papilla and the seminal vesicle); free-living *Notoplana atomata* (O. F. Müller, 1776)

8b Penis stylet nearly twice as long as the prostatic vesicle; inhabiting shells occupied by hermit crabs. *Notoplana inquilina* Hyman, 1955

9a With well developed spermiducal bulbs (dilations of the sperm ducts) 10

9b Without spermiducal bulbs 11

10a Gonopore single, conspicuous; penis papilla large, tonguelike (often protruding from the gonopore in preserved specimens) *Freemania litoricola* (Heath & McGregor, 1912)

10b Male and female gonopores separate, inconspicuous; penis papilla very small *Leptoplana chloranota* (Boone, 1929)

11a Color red or wine, or marked with a large blotch of red 12

11b Color grayish or brownish, nearly uniform or mottled, but definitely not red or wine, and not marked with a large blotch of red 13

12a Marked with a red saddlelike blotch on the central part of the dorsal surface, elsewhere gray; male and female genital pores separate *Notoplana sanguinea* Freeman, 1933

12b Color red or wine throughout; gonopore single *Notoplana celeris* Freeman, 1933

13a Prostatic vesicle tubular *Leptoplana vesiculata* Hyman, 1939

13b Prostatic vesicle ovoid 14

14a Prostatic vesicle smaller than the seminal vesicle; penis papilla small, conical, not easily identified in entire specimens; not more or less obligatorily associated with oysters 15

14b Prostatic vesicle larger than the seminal vesicle; penis papilla large, bulbous, muscular, readily identified in entire specimens; regularly associated with oysters *(Ostrea lurida)*, and a predator on younger individuals *Pseudostylochus ostreophagus* Hyman, 1955

15a Male and female genital pores very close together, the distance between them much shorter than the length of Lang's vesicle; capable of swimming effectively by fluttering movements of the entire body *Notoplana natans* Freeman, 1933

15b Male and female genital pores rather widely separated, the distance between them exceeding the length of Lang's vesicle; not capable of swimming effectively *Notoplana sanjuania* Freeman, 1933

16a With a pair of marginal tentacles at the anterior end 18

16b Without marginal tentacles at the anterior end 17

17a Mouth located at the center of the anterior margin; male gonopore separate from the mouth *Acerotisa alba* (Freeman, 1933)

17b Mouth located posterior to the brain; male antrum opening into the mouth *Stylostomum sanjuania* Holleman, 1972

18a Background color yellowish pink or salmon, with minute white specks, or uniformly orange-red *Eurylepta aurantiaca* Heath & McGregor, 1912

18b Background color cream, light tan, brown, or gray, without any distinctly pink, orange, or red tones 19

19a Background color cream or light tan, marked on the dorsal surface with conspicuous dark reddish brown spots *Eurylepta leoparda* Freeman, 1933

19b Background color light brown to gray, mottled with dark brown *Pseudoceros canadensis* Hyman, 1953

Bock, S. 1925. Papers from Dr. Th. Mortensen's Pacific expedition 1914-1918. XXVII. Planarians. Pt. IV. New stylochids. Vidensk. Medd. Dansk Naturh. Foren., 79:97-198.

Boone, E. 1929. Five new polyclads from the California coast. Ann. Mag. Nat. Hist., ser. 10, 3:33-46.

Freeman, D. 1933. The polyclads of the San Juan region of Puget Sound. Trans. Amer. Micr. Soc., 52:107-46.

Heath, H., & E. A. McGregor. 1912. New polyclads from Monterey Bay, California. Proc. Acad. Nat. Sci. Philadelphia, 64:455-87.

Holleman, J. J. 1972. Marine turbellarians of the Pacific coast I. Proc. Biol. Soc. Washington, 85:405-7.

Hyman, L. H. 1939. New species of flatworms from North, Central and South America. Proc. U. S. Nat. Mus., 86:419-39.
------. 1939. Some polyclads of the New England coast, especially of the Woods Hole region. Biol. Bull., 76:127-52.
------. 1953. The polyclad flatworms of the Pacific coast of North America. Bull. Amer. Mus. Nat. Hist., 100:265-392.
------. 1955. The polyclad flatworms of the Pacific coast of North America: additions and corrections. Amer. Mus. Novit., No. 1704. 11 pp.
------. 1959. Some Turbellaria from the coast of California. Amer. Mus. Novit., No. 1943. 17 pp.

SUBCLASS EULECITHOPHORA

ORDER NEORHABDOCOELIDA ("RHABDOCOELS")

Free-living Species

Although these are regularly encountered in almost all habitats, they have been little studied in our region. This is distressing, for we must have no fewer than 50 species. It is possible that some of the rhabdocoels reported by Karling (1962a, 1962b, 1967) from California occur in our area. However, one must be cautious in using these papers, because undescribed species superficially similar to those studied by Karling might be incorrectly identified.

Karling, T. G. 1962a. Marine Turbellaria from the Pacific coast of North America. I. Plagiostomidae. Arkiv för Zoologi, ser. 2, 15:113-41.
------. 1962b. Marine Turbellaria from the Pacific coast of North America. II. Pseudostomidae and Cylindrostomidae. Arkiv för Zoologi, ser. 2, 15:181-209.
------. 1967. On the genus *Promesostoma* (Turbellaria), with descriptions of four new species from Scandinavia and California. Sarsia, 29:257-68.

Family Umagillidae

The species known to occur in the digestive tracts of sipunculans and echinoderms in our region are arranged according to host.

Sipuncula
- *Themiste pyroides* (Chamberlin)
  - *Collastoma pacifica* Kozloff, 1953

Echinodermata
- Crinoidea
  - *Florometra serratissima* (A. H. Clark)
    - *Desmote inops* Kozloff, 1965
    - *Fallacohospes inchoatus* Kozloff, 1965
- Echinoidea
  - *Strongylocentrotus purpuratus* (Stimpson)
    - *Syndisyrinx franciscanus* Lehman, 1946
  - *Strongylocentrotus pallidus* (G. O. Sars)
    - *Syndisyrinx franciscanus* Lehman, 1946
  - *Strongylocentrotus droebachiensis* (Müller)
    - *Syndisyrinx franciscanus* Lehman, 1946; regularly present

Syndesmis sp. (possibly *S. echinorum* Francois, 1886); found only occasionally; distinguished from the foregoing by its brownish [instead of reddish] color)

Strongylocentrotus franciscanus (Stimpson)

Syndisyrinx franciscanus Lehman, 1946

Allocentrotus fragilis (Jackson)

Syndesmis sp.

Dendraster excentricus (Eschscholtz)

Syndesmis dendrastorum Stunkard & Corliss, 1951

Holothuroidea

Parastichopus californicus (Stimpson)

Ozametra sp.

Kozloff, E. N. 1953. *Collastoma pacifica* sp. nov., a rhabdocoel turbellarian from the gut of *Dendrostoma pyroides* Chamberlin. J. Parasit., 39: 336-40.

------. 1965. *Desmote inops* sp. n. and *Fallacohospes inchoatus* gen. and sp. n., umagillid rhabdocoels from the intestine of the crinoid *Florometra serratissima* (A. H. Clark). J. Parasit., 51:305-12.

Lehman, H. E. 1946. A histological study of *Syndisyrinx franciscanus* gen. et sp. nov., an endoparasitic rhabdocoel of the sea urchin, *Strongylocentrotus franciscanus*. Biol. Bull., 91:295-311.

Stunkard, H. W., & J. O. Corliss. 1951. A new species of *Syndesmis* and a revision of the family Umagillidae Wahl, 1910 (Turbellaria: Rhabdocoela). Biol. Bull., 101:319-34.

ORDER PROSERIATA ("ALLOEOCOELS")

Many species of this group occur in our fauna. A few of them have been described in the papers of Ax (1966) and Ax & Ax (1967, 1969). Two papers of Karling (1964, 1966) on California species may be helpful in assigning some others to genus, at least.

Family Otoplanidae

Subfamily Otoplaninae

Americanaplana fernaldi Ax & Ax, 1967 (intertidal, in a mixture of coarse sand and gravel)

Pluribursaeplana illgi Ax & Ax, 1967 (in coarse sediment with stones, at depths of about 50 m)

Orthoplana kohni Ax & Ax, 1967 (intertidal, in coarse sand)

Itaspiella armata Ax, 1951, subsp. *magna* Ax & Ax, 1967 (intertidal, in coarse sand and gravel)

Notocaryoplanella glandulosa Ax, 1951 (intertidal, in coarse and medium sand or in a mixture of coarse sand and gravel)

Subfamily Parotoplaninae

Polyrhabdoplana posttestis Ax & Ax, 1967 (intertidal, in a mixture of coarse sand and gravel)

Parotoplana pacifica Ax & Ax, 1967 (in sediment, at depths of about 50 meters)

Philosyrtis sanjuanensis Ax & Ax, 1967 (intertidal, in fine to medium sand)

Ax, P. 1966. Die Bedeutung der interstitiellen Sandfauna für allgemeine Probleme der Systematik, Ökologie und Biologie. Veröffentl. Inst.

Meeresforsch. Bremerhaven, Sonderbd. 2:15-65.
Ax, P., & R. Ax. 1967. Turbellaria Proseriata vor der Pazifikküste der USA (Washington). I. Otoplanidae. Zeitschr. Morph. Tiere, 61:215-54.
------. 1969. Eine Chorda intestinalis bei Turbellarien (*Nematoplana nigrocapitula* Ax) als Modell für die Evolution der Chorda dorsalis. Akad. Wiss. Lit. Mainz, Abhandl. Math. Naturh. Kl., Jahrg. 1969:133-49.
Karling, T. G. 1964. Marine Turbellaria from the Pacific coast of North America. III. Otoplanidae. Arkiv för Zoologi, ser. 2, 16:527-41.
------. 1966. Marine Turbellaria from the Pacific coast of North America. IV. Coelogynoporidae and Monocelididae. Arkiv för Zoologi, ser. 2, 18: 493-528.

ORDER TRICLADIDA

Four species have been found in our general area, although 2 (*D. opisthopora* and *P. psammophilus*) have not been reported north of Oregon.

Nexilis epichitonius Holleman & Hand, 1962
unusual in lacking a vagina, and in having the ovovitelline ducts entering the inflated base of the penis, just posterior to the seminal vesicle; commensal with chitons and mussels, and sometimes found on the glass of aquaria containing these molluscs
Oregoniplana opisthopora Holmquist & Karling, 1972
in the sandy substrates of *Zostera* beds, and also on rocks
Pacificides psammophilus Holmquist & Karling, 1972
Procerodes pacifica Hyman, 1954
in coarse gravel in the intertidal zone

Holleman, J. J. 1972. Marine turbellarians of the Pacific coast I. Proc. Biol. Soc. Washington, 85:405-07.
Holleman, J. J., & C. Hand. 1962. A new species, genus, and family of marine flatworms (Turbellaria: Tricladida, Maricola) commensal with mollusks. Veliger, 5:20-2.
Holmquist, C., & T. G. Karling. 1972. Two new species of interstitial marine triclads from the North American Pacific coast, with comments on evolutionary trends and systematics in Tricladida. Zool. Scripta, 1: 175-84.
Hyman, L. H. 1954. A new marine triclad from the coast of California. Amer. Mus. Novit., no. 1679. 5 pp.

PHYLUM ORTHONECTIDA

Only 2 species of orthonectids have been reported from our region. However, it is probable that if large numbers of certain prospective hosts (especially turbellarians, nemerteans, polychaete annelids, and pelecypod molluscs) were examined systematically, others would be found.

Rhopalura ophiocomae Giard, 1877
In the gonads of the ophiuroid *Amphipholis squamata*. The incidence of parasitized hosts is generally low. Until recently, the only records of

parasitism of *A. squamata* by *R. ophiocomae* have been on the Atlantic coast of France, Mediterranean coasts of France and Italy, and the British Isles. It is likely that the parasite will be found almost throughout the very extensive range of its host. For a detailed description, and a complete bibliography of original work on this species, see Kozloff (1969).

Ciliocincta sabellariae Kozloff, 1965
In the body wall of the polychaete *Sabellaria cementarium.* The incidence of parasitized hosts is very low. For a detailed description, see Kozloff (1971).

Kozloff, E. N. 1965. *Ciliocincta sabellariae* gen. and sp. n., an orthonectid mesozoan from the polychaete *Sabellaria cementarium* Moore. J. Parasit., 51:37-44.

------. 1969. Morphology of the orthonectid *Rhopalura ophiocomae.* J. Parasit., 55:171-95.

------. 1971. Morphology of the orthonectid *Ciliocincta sabellariae.* J. Parasit., 57:585-97.

PHYLUM DICYEMIDA

F. G. Hochberg
Santa Barbara Museum of Natural History

The 4 species described from cephalopod molluscs in our area are arranged according to host.

Rossia pacifica Berry
- *Dicyemennea brevicephaloides* Bogolepova-Dobrokhotova, 1962
  - in the branchial heart coelom
  - vermiform stage: length to 3 mm; calotte flattened; 23 somatic cells
  - infusoriform stage: length 30 μ; without refringent bodies
- *Dicyemennea filiformis* Bogolepova-Dobrokhotova, 1962 *(=D. parva* Hoffman, 1965)
  - in the renal coelom
  - vermiform stage: length to 1.5 mm; calotte elongated and narrow; 24 to 26 somatic cells
  - infusoriform stage: length 25 μ; without refringent bodies

Octopus dofleini martini Pickford *(=O. apollyon,* as used by McConnaughey, 1957)
- *Conocyema deca* McConnaughey, 1957
  - in the renal coelom
  - vermiform stage: length not exceeding 1 mm; calotte flattened; axial cell terminating against bases of propolar cells; 23 or 24 somatic cells
  - infusoriform stage: length to 35 to 40 μ; with refringent bodies
- *Dicyemennea abreida* McConnaughey, 1957
  - in the renal coelom
  - vermiform stage: length to 3 mm; calotte rounded; axial cell terminating between bases of propolar cells; 28 to 30 somatic cells
  - infusoriform stage: length 35 μ; with refringent bodies

Bogolepova-Dobrokhotova, I. I. 1962 [Dicyemidae of the Far-Eastern Seas. II. New species of the genus Dicyemennea.] Zool. Zhurnal SSSR, 41:503-18. (in Russian)
Hoffman, E. G. 1965. Mesozoa of the sepiolid, *Rossia pacifica* (Berry). J. Parasit., 51:313-20.
McConnaughey, B. H. 1957. Two new mesozoans from the Pacific Northwest. J. Parasit., 43:358-64.

PHYLUM NEMERTEA

The key does not include *Carcinonemertes epialti* Coe, 1902 (order Hoplonemertea). It is occasionally found in egg masses of crabs *(Cancer magister, Hemigrapsus oregonensis, Pugettia producta)*.

1a Body leechlike; posterior end provided with a muscularized disc for attachment; commensal in mantle cavity of bivalve molluscs *(Siliqua patula, Macoma nasuta, Macoma secta, Tresus capax,* and others) (Class Enopla: Order Bdellonemertea) *Malacobdella* spp.
1b Body not leechlike; without an attachment disc at posterior end; free-living
2a Mouth decidedly ventral in position, behind the brain, and separate from the proboscis pore; proboscis not armed with stylets (to determine whether stylets are present, flatten the animal [if it is small and transparent] or an isolated proboscis carefully between microscope slides and examine it with transmitted light) 3
2b Mouth essentially at the anterior end of the body, in front of the brain, united with the proboscis pore; proboscis armed with 1 or more stylets (Class Enopla: Order Hoplonemertea) 17
3a With distinct longitudinal cephalic grooves on both sides of the head (Class Anopla: Order Heteronemertea) 9
3b Without longitudinal cephalic grooves at the sides of the head (Class Anopla: Order Paleonemertea) 4
4a Body marked with distinct white rings or longitudinal lines, or both 5
4b Coloration nearly homogeneous, without definite markings 6
5a Ground color deep red, interrupted by nearly evenly spaced white rings (the spacing is less regular in the anterior region) *Tubulanus albocinctus* (Coe, 1904)
5b Ground color ranging from orange-brown to chocolate, interrupted by nearly evenly spaced white rings and also by 5 or 6 longitudinal white lines (one mid-dorsal, 2 lateral on each side; if a sixth line is present, it is mid-ventral, and usually rather weak) *Tubulanus sexlineatus* (Griffin, 1898)
6a Ground color whitish or cream; darker pigmentation, if present, not dominating the background color; length of large specimens not exceeding 50 cm 7
6b Ground color orange or yellow; darker pigmentation, if present, dominating the background color; length of large specimens attaining 2 m 8
7a Small and slender, not exceeding 3 cm; without dark pigmentation (the eggs are cream-colored, and because of this gravid females may show a mid-dorsal and mid-ventral yellow or pale orange band); head region translucent; body not conspicuously dilated posteriorly and not show-

ing distinctly paired intestinal diverticula
Tubulanus pellucidus (Coe, 1895)

7b Larger, sometimes attaining 50 cm; background color nearly white, mottled with small patches of brownish pigment; head milk white, not translucent; most of body behind the head conspicuously flattened, showing paired intestinal diverticula (in specimens with well developed gonads, these lie transversely between the diverticula, and confer an orange color [in the male] or reddish color [in the female] on the flattened part of the body) *Carinoma mutabilis* Griffin, 1898

8a Color uniformly bright orange or red; head not capable of being completely withdrawn into rest of body; no eyespots on head
Tubulanus polymorphus Renier, 1804

8b Ground color yellow, sprinkled with small irregular dark spots; head capable of being completely withdrawn into rest of body; tip of head with 2 groups of numerous minute eyespots
Baseodiscus princeps (Coe, 1901)

9a Caudal cirrus present (caution!--it may be missing temporarily); proboscis sheath, if it can be seen, usually nearly as long as the body 11

9b Caudal cirrus absent; proboscis sheath, if it can be seen, usually much shorter than the body 10

10a General coloration pink, pinkish red, or rosy, sometimes with a tinge of blue (preserved specimens may show a bluish tinge also); tip of head white both above and below, thereby contrasting sharply with the rest of the body; generally 2 to 4 eyespots (rarely 6 to 8) on either side of head *Lineus rubescens* Coe, 1904

10b General coloration dusky, brownish green, dark brown, or reddish brown (a faint, light longitudinal line may be present on each lateral margin, and there may be a number of faint rings encircling the body), commonly paler below; coloration of tip of head not conspicuously different from rest of body; 4 to 8 eyespots on either side of head
Lineus vegetus Coe, 1931

11a Body rather soft, slender, without thinner margins, and not adapted for swimming 12

11b Body firm, flattened in the intestinal region, with thin lateral margins, and thereby adapted for swimming 14

12a With transverse white lines or white rings encircling the body; background coloration of dorsal surface dark 13

12b Without transverse white lines or white rings encircling the body; coloration of dorsal surface ranging from salmon to light brown, but never dark *Micrura alaskensis* Coe, 1901

13a Tip of head bright orange dorsally; rest of dorsal surface purplish brown or reddish brown, with regularly spaced transverse white lines; ventral surface white *Micrura verrilli* Coe, 1901

13b Tip and sides of head white; rest of dorsal surface some shade of brown, brownish black, or dark gray, with irregularly spaced transverse white lines that completely encircle the body; ventral surface about the same color as the dorsal surface, but lighter
Micrura wilsoni (Coe, 1901)

14a Tip of the head, or a substantial part of the head, white or yellowish both above and below, and therefore contrasting sharply with the rest of the body, which is dark brown, reddish purple, or red 15

14b Tip of the head not contrasting sharply with the rest of the body, which may be relatively light or very dark 16

15a Body deep red; tip of head white or yellowish
Cerebratulus montgomeryi Coe, 1901

15b Body dark brown, reddish purple, or nearly black; much of the head (to a level well behind the middle of the cephalic grooves) white
Cerebratulus albifrons Coe, 1901

16a Color ranging from pale yellow or light gray to dark gray or chocolate; lateral margins may be pale, but not almost white; lateral nerve cords often distinctly visible as reddish lines near the margins
Cerebratulus californiensis Coe, 1905

16b Color generally brown, gray, or olive; lateral margins often almost white; lateral nerve cords not distinctly visible as reddish lines near the margins *Cerebratulus marginatus* Renier, 1804
(The last 2 species of Cerebratulus are difficult to distinguish from one another and from *C. herculeus* Coe, 1901 and *C. occidentalis* Coe, 1901, also reported from this region. The best source of information on this complex is Coe, 1905.)

17a Proboscis sheath not more than three-fourths the length of the body 18

17b Proboscis sheath nearly as long as the body 19

18a General coloration dark green above, yellowish below; proboscis sheath, when evident, less than half as long as body; proboscis generally with 2 (occasionally with 3) pouches of accessory stylets; stylets curved, without a braided appearance
Emplectonema gracile (Johnston, 1837)

18b General coloration dark brown or purplish brown, but lighter midventrally; proboscis sheath, when evident, one-half to three-fourths the length of the body; proboscis with 2, 4, or 6 pouches of accessory stylets; stylets straight, with a braided appearance due to spiral sculpturings *Paranemertes peregrina* Coe, 1901

19a Small worms rarely longer than 1.5 or 2 cm at maturity; generally with 4 conspicuous eyespots that form a quadrangle on the head; intestinal caecum with short anterior branches, or these are absent altogether; intestinal diverticula unbranched *Tetrastemma* spp.

19b Generally longer than 2 cm at maturity; eyespots numerous, unless there is only 1 pair or they are absent altogether, never 4; intestinal caecum with long anterior branches; intestinal diverticula branched
20

20a Eyespots extending along lateral nerve cords for some distance posterior to the brain; proximal end of basal portion of central stylet abruptly truncated *Zygonemertes virescens* (Verrill, 1879)

20b Eyespots not extending posterior to the brain; proximal end of basal portion of central stylet rounded 21

21a Body slender; general coloration whitish, sometimes tinged lightly with red or yellow spots 22

21b Body rather stout; general coloration red, brown, purplish, orange, or deep yellow, the pigment sometimes concentrated in conspicuous spots
23

22a With 2 or 3 pouches of accessory stylets
Amphiporus imparispinosus Griffin, 1898

22b With 6 to 12 pouches of accessory stylets
Amphiporus formidabilis Griffin, 1898

23a Ground color of head much lighter than rest of dorsal surface, and marked with a pair of conspicuous dark, angular or oval blotches; central stylet about twice as long as its basal portion 24

23b Ground color of head essentially the same as that of the dorsal sur-

face in general, and without a pair of conspicuous blotches; central stylet approximately the same length as its basal portion
Amphiporus rubellus Coe, 1905

24a General coloration of dorsal surface nearly homogeneous, reddish or brown; dark blotches on head similarly uniform
Amphiporus bimaculatus Coe, 1901

24b Dark brown coloration of dorsal surface concentrated in irregular blotches, the yellowish or whitish ground color being evident between the blotches; dark blotches on head similarly heterogeneous
Amphiporus punctulatus Coe, 1905

Coe, W. R. 1901. Papers from the Harriman Alaska Expedition. XX. The nemerteans of the expedition. Proc. Wash. Acad. Sci., 3:1-110. (Reprinted in Harriman Alaska Expedition, 11:1-110 [1904])

------. 1904. Nemerteans of the Pacific coast of North America. Part II. Harriman Alaska Expedition, 11:111-220.

------. 1905. Nemerteans of the west and northwest coast of America. Bull. Mus. Comp. Zool. Harvard, 47:1-319.

------. 1940. Revision of the nemertean fauna of the Pacific coast of North, Central, and northern South America. Allan Hancock Pacific Expeditions, 2:247-323.

Correa, D. D. 1964. Nemerteans from California and Oregon. Proc. Calif. Acad. Sci., ser. 4, 31:515-58.

Humes, A. G. 1942. The morphology, taxonomy, and bionomics of the nemertean genus *Carcinonemertes*. Illinois Biol. Monogr., 18(4):1-105.

PHYLUM ROTIFERA

There appear to be no papers concerned with the fauna of marine rotifers of our region. However, the key published by Thane-Fenchel (1968) will enable one to identify the genera known to occur in marine and brackish-water habitats in various parts of the world; it is accompanied by a useful bibliography.

A species of the rather bizarre genus *Seison* is found on the body surface of *Nebalia pugettensis*, a leptostracan crustacean that is sometimes common in pools where algae are decomposing. This species of *Seison* may be one of those reported from leptostracans in Europe, but there is a good chance that it is undescribed.

Thane-Fenchel, A. 1968. A simple key to the genera of marine and brackish-water rotifers. Ophelia, 5:299-311.

PHYLUM GASTROTRICHA

Only a few species of gastrotrichs have been reported from Puget Sound and the San Juan Archipelago. There are no published records for the outer coast of our region. As a key would very likely lead to incorrect identification of undescribed species, it seems best to provide only a checklist and a bibliography of pertinent literature.

ORDER CHAETONOTIDA

Chaetonotus testiculophorus Hummon, 1966 (intertidal, in sand)
Musellifer sublittoralis Hummon, 1969 (subtidal, in epibenthic sediment)

ORDER MACRODASYIDA

Macrodasys cunctatus Wieser, 1957 (intertidal, in sand)
Tetranchyroderma pugetensis Wieser, 1957 (high intertidal, in coarse sand)
Paraturbanella intermedia Wieser, 1957 (low intertidal, in sand)
Turbanella cornuta Remane, 1925 (intertidal, in sand)
Turbanella mustela Wieser, 1957 (high intertidal, in sand)

Hummon, W. D. 1966. Morphology, life history, and significance of the marine gastrotrich, *Chaetonotus testiculophorus* n. sp. Trans. Amer. Micr. Soc., 85:450-57.

------. 1969. *Musellifer sublittoralis,* a new genus and species of Gastrotricha from the San Juan Archipelago, Washington. Trans. Amer. Micr. Soc., 88:283-86.

Remane, A. 1925. Neue aberrante Gastrotrichen II: *Turbanella cornuta* nov. spec. und *T. hyalina* M. Schultze, 1853. Zool. Anz., 64:309-14.

------. 1936. Gastrotricha und Kinorhyncha. *In* Bronn, H. G. (ed.), Klassen und Ordnungen des Tierreichs, 4, 2:1-242.

Wieser, W. 1957. Gastrotricha Macrodasyoidea from the intertidal of Puget Sound. Trans. Amer. Micr. Soc., 76:372-81.

------. 1959. Free-living nematodes and other small invertebrates of Puget Sound beaches. Univ. Wash. Publ. Biol., 19. x + 179 pp.

PHYLUM KINORHYNCHA

John C. Boykin
Seattle, Washington

Of the several species of kinorhynchs known to occur in this region, all but one are subtidal, living near the surface of soft mud. The only intertidal species, sometimes common in the sediment on muddy sand, belongs to the genus *Echinoderes,* but is undescribed (couplet 4a).

1a Exoskeleton of segment 3 (first trunk segment) either a closed ring or consisting of 2 half rings; generally circular or oval in transverse section; lateral and dorsal setae present on trunk and usually longer than one-third the trunk diameter (Order Cyclorhagida) 2

1b Exoskeleton of segment 3 with a large, strongly arched dorsal tergite, a single midventral sternite, and 2 lateroventral sternites; triangular in transverse section; trunk setae present, but shorter than one-fourth the trunk diameter (Order Homalorhagida) 6

2a Exoskeleton of both segments 3 and 4 (first and second trunk segments) consisting of closed rings; median terminal spine present only in colorless juvenile forms no longer than about 250 μ (Family Echinoderidae) 3

2b Exoskeleton of segment 4 (second trunk segment) divided into a dorsal tergite and 2 ventral sternites; median terminal spine (as well as lateral terminal spines) present in adults (Family Centroderidae)
Campyloderes sp. [possibly *C. multispinosus* (McIntyre, 1962)]

3a Length less than 300 μ; exoskeleton colorless; smaller specimens may have a median terminal spine as well as a pair of lateral terminal spines *Echinoderes*, juveniles
3b Length greater than 300 μ; exoskeleton yellow or yellow-brown; with a pair of well developed lateral terminal spines, but without a median terminal spine (*Echinoderes*, adults) 4

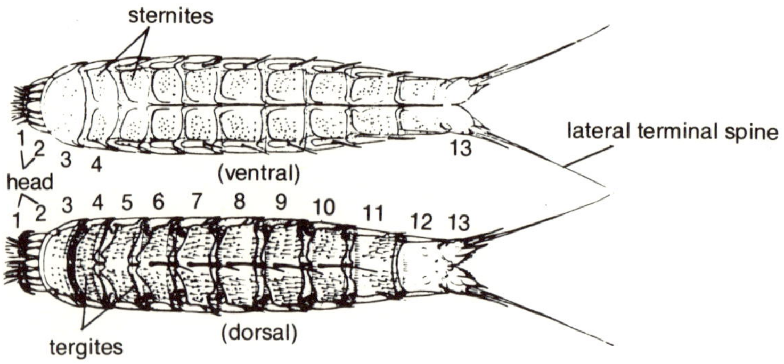

Echinoderes

4a With a red eyespot on each side of the head and segmental accumulations of red pigment along the mid-dorsal line (locally abundant intertidally, in sediment on muddy sand) *Echinoderes* sp. (undescribed) This species is close to *E. dujardini* Claparède, 1863 and to *E. worthingi* Zelinka, 1928 (see Kozloff, 1972), but appears to be distinct from both.
4b Without red eyespots on the head or accumulations of red pigment along the mid-dorsal line 5
5a Length greater than 390 μ; lateral terminal spines 135 to 180 μ long *Echinoderes pennaki* Higgins, 1960
5b Length 310 to 365 μ; lateral terminal spines 50 to 65 μ *Echinoderes* sp. (undescribed)
6a Without movable terminal spines on segment 13 (last trunk segment) (genus *Trachydemus*) 7
6b With a single median terminal spine or a pair of lateral terminal spines on segment 13 (genus *Pycnophyes*) 9
7a Length less than 500 μ; exoskeleton colorless juveniles of *Trachydemus*
7b Length greater than 500 μ; exoskeleton yellow to yellow-brown in color 8
8a Segment 3 (first trunk segment) with a pair of dorsolateral setae; segment 13 (last trunk segment) with 2 pairs of lateral setae (these setae are only 10-12 μ long, and easily broken in handling) *Trachydemus cataphractus* Higgins, 1961
8b Segment 3 without dorsolateral setae; segment 13 with one pair of lateral setae *Trachydemus ilyocryptus* Higgins, 1961
9a Length less than 450 μ; exoskeleton colorless; with a single median terminal spine younger juveniles of *Pycnophyes*
9b Length greater then 450 μ; with a pair of lateral terminal spines 10
10a Exoskeleton colorless older juveniles of *Pycnophyes*

10b Exoskeleton yellow to yellow-brown
adult *Pycnophyes sanjuanensis* Higgins, 1961

Higgins, R. P. 1960. A new species of *Echinoderes* (Kinorhyncha) from Puget Sound. Trans. Amer. Micr. Soc., 79:85-91.

------. 1961. Three new homalorhage kinorhynchs from the San Juan Archipelago, Washington. J. Elisha Mitchell Sci. Soc., 77:81-8.

Kozloff, E. N. 1972. Some aspects of development in *Echinoderes* (Kinorhyncha). Trans. Amer. Micr. Soc., 91:119-30.

McIntyre, A. D. 1962. The class Kinorhyncha (Echinoderida) in British waters. J. Mar. Biol. Assoc. U. K., 42:503-4.

Zelinka, C. 1928. Monographie der Echinodera. Leipzig: Wilhelm Engelmann. 396 pp.

PHYLUM NEMATODA

Nematodes are abundant in sand and mud, and in the sediment that accumulates on sessile invertebrates, algal growths, and other substrates. There are also many species living as parasites or commensals in marine vertebrates and invertebrates.

The only comprehensive work on free-living nematodes in this area is that of Wieser (1959). He found about 80 species, distributed over 16 families. Most of the species recorded by Wieser were new. As his study was confined to a few beaches in Puget Sound, it is obvious that the fauna of nematodes in our region must be very large. Wieser provided a helpful bibliography on free-living nematodes up to 1951. Some additional publications dealing with nematodes of the west coast are cited below; the bibliographies in these should prove useful.

The literature on parasitic and commensal nematodes is understandably large. There are no published general catalogues of species found in marine fishes, mammals, birds, and invertebrates of the Northwest.

Allgén, C. A. 1947. Papers from Dr. Th. Mortensen's Pacific Expedition 1914-1916. LXXXV. West American marine nematodes. Videnskab. Medd. Dansk Naturh. Foren. Kφbenhavn, 110:65-219.

------. 1951. Papers from Dr. Th. Mortensen's Pacific Expedition 1914-1916. LXXXVI. Pacific freeliving marine nematodes. Videnskab. Medd. Dansk Naturh. Foren. Kφbenhavn, 113:263-411.

Chitwood, B. G. 1960. A preliminary contribution to the marine nemas (Adenophorea) of northern California. Trans. Amer. Micr. Soc., 79:347-84.

Hope, W. D. 1967. Free-living marine nematodes of the genera *Pseudocella* Filipjev, 1927, *Thoracostoma* Marion, 1870, and *Deontostoma* Filipjev, 1916 (Nematoda: Leptosomatidae) from the west coast of North America. Trans. Amer. Micr. Soc., 86:307-34.

Hope, W. D., & D. G. Murphy. 1972. A taxonomic hierarchy and checklist of the genera and higher taxa of marine nematodes. Smithsonian Contrib. Zool., no. 137. 100 pp.

Murphy, D. G. 1965. Free-living marine nematodes II. *Thoracostoma pacifica* n. sp. from the coast of Oregon. Proc. Helminthol. Soc. Wash., 32:106-09.

------. 1965. *Thoracostoma washingtoniensis* n. sp., eine Meeresnematode aus dem Pazifischen Kustenbereich vor Washington. Abhandl. Verhandl. Naturwiss. Vereins Hamburg, n. F., 9:211-16.

Wieser, W. 1959. Free-living nematodes and other small invertebrates of Puget Sound beaches. Univ. Wash. Publ. Biol., 19. x + 179 pp.

PHYLUM ENTOPROCTA

1a Zooids separate from one another, not connected by stolons; generally attached to other organisms (Family Loxosomatidae) 2
1b Zooids in colonies, connected by stolons; free-living or attached to other organisms (Family Pedicellinidae) 3
2a Base of stalk differentiated into a muscular disk by which the animal is attached; attachment not permanent, the animal capable of moving about to some extent *Loxosoma* sp.
2b Base of stalk not differentiated into a muscular disk; permanently attached to the skin of the polychaete *Diopatra ornata* (Onuphidae) *Loxosomella* sp.
3a Stalks thick, without muscular enlargements where they join the stolon 4
3b Stalks thin, with muscular enlargements where they join the stolon (and sometimes with 1 or more muscular enlargements between the base and the calyx) 5
4a Ring of tentacles in a plane that is decidedly oblique to the long axis of the stalk; both stalk and calyx strongly muscularized *Myosoma spinosa* Robertson, 1900
4b Ring of tentacles in a plane that is at right angles to the long axis of the stalk; neither stalk nor calyx strongly muscularized *Pedicellina cernua* (Pallas, 1771)
5a Stalks with 1 or more muscular enlargements (in addition to those at their bases) that enable them to bend sharply 6
5b Stalks without muscular enlargements (other than those at their bases) 7
6a Stalks with pores, and branching by producing new individuals at the points where the muscular enlargements are located *Barentsia ramosa* (Robertson, 1900)
6b Stalks without pores, and rarely branching *Barentsia benedeni* (Foettinger, 1887)
7a Stalks with numerous pores; height of individual zooids up to about 4 mm *Barentsia discreta* (Busk, 1886)
7b Stalks with few or no pores; height of individual zooids not often exceeding 1 mm *Barentsia gracilis* (M. Sars, 1835)

Mariscal, R. N. 1965. The adult and larval morphology and life history of the entoproct *Barentsia gracilis* (M. Sars, 1835). J. Morphol., 116: 311-38. (Concerns *B. benedeni.*)

Osburn, R. C. 1953. Bryozoa of the Pacific coast of America. Part 3. Cyclostomata, Ctenostomata, Entoprocta, and addenda. Allan Hancock Pacific Exped., 14:613-841.

Robertson, A. 1900. Studies in Pacific coast Entoprocta. Proc. Calif. Acad. Sci., ser. 3, Zool., 2:323-48.

PHYLUM MOLLUSCA

The following general references deal with more than 1 class of molluscs represented in our fauna. References that are concerned with individual classes are listed after the keys and other material that pertain to these groups.

Bernard, F. R. 1970. A distributional checklist of the marine molluscs of British Columbia: based on faunistic surveys since 1950. Syesis, 3:75-94.

Burch, J. Q. (ed.). 1944-46. Distributional list of the west American marine molluscs from San Diego, California, to the Polar Sea. Minutes Conchol. Club So. Calif., nos. 45-63. (mimeographed)

Dall, W. H. 1921. Summary of the marine shell-bearing mollusks of the northwest coast of America, from San Diego, California, to the Polar Sea, mostly contained in the collection of the United States National Museum, with illustrations of hitherto unfigured species. Bull. 112, U. S. Nat. Mus. iii + 217 pp.

Keen, A. M., & E. Coan. 1974. Marine Molluscan Genera of Western North America: An Illustrated Key. 2nd ed. Stanford, California: Stanford University Press. vi + 208 pp.

Oldroyd, I. S. 1924. Marine Shells of Puget Sound and vicinity. Publ. Puget Sound Biol. Station, 4:1-272.

------. 1925-27. The marine shells of the west coast of North America. Stanford Univ. Publ. Geol. Sci., 1:1-247; 2:1-941.

Quayle, D. B. 1964. Distribution of introduced marine mollusca in British Columbia waters. J. Fish. Res. Bd. Canada, 21:1155-81.

Rice, T. 1971. Marine Shells of the Pacific Northwest. Edmonds, Washington: Ellison Industries. 102 pp.

CLASS AMPHINEURA

SUBCLASS APLACOPHORA

A species of *Chaetoderma* (Chaetodermatidae) is occasionally collected by dredging on soft substrates. *Proneomenia ?sluiteri* Hubrecht, 1880 (Neomeniidae) is reported subtidally from the area north of Vancouver Island.

SUBCLASS POLYPLACOPHORA

The key covers almost all species of chitons likely to be found intertidally or by dredging down to depths of about 100 m. However, the following species are not included:

Mopalia cirrata Berry, 1919
Mopalia laevior Pilsbry, 1918
Mopalia sinuata Carpenter, 1864
Mopalia swanii Carpenter, 1864
Leptochiton cancellatus (Sowerby, 1839)

The mopalias listed above are difficult to separate from one another as well as from certain of the 5 species that are in the key. Moreover, there may be some undescribed species of *Mopalia* in this region. In any case, it seems best not to try to assimilate any of these species until specimens positively identified by a specialist can be carefully examined with a view to finding some distinctive features by which they can be distinguished from species that are included. The references cited in the bibliography will be of little help in dealing with the less easily recognizable chitons of our area. Monographic studies of several genera are sorely needed.

1a All 8 valves completely covered by the tough, reddish brown girdle; length sometimes exceeding 20 cm *Cryptochiton stelleri* (Middendorff, 1846)
1b All 8 valves visible; length rarely exceeding 10 cm 2
2a Girdle smooth, without hairs, bristles, spines, spicules, scales, or obvious granulations 3
2b Girdle with hairs, bristles, spines, spicules, scales, or obvious granulations 5
3a Girdle black, so extensive as to nearly cover the valves; valves without zigzag markings in various bright colors; length frequently exceeding 6 cm *Katharina tunicata* (Wood, 1815)
3b Girdle not black, and not so extensive as to nearly cover the valves; valves with prominent zigzag markings of white, yellow, pink, red, purple, and other colors; length not often exceeding 5 cm 4
4a Some of the valves with transverse light lines near the midline; girdle usually with subdued green blotches *Tonicella insignis* Reeve, 1847
4b Valves without transverse light lines near the midline; girdle usually with conspicuous whitish to orange blotches *Tonicella lineata* (Wood, 1815)
5a Girdle with hairs, flexible bristles, stiff spines, or microscopic, glassy spicules (if spicules are present, there may also be fine granules) 6
5b Girdle without hairs, bristles, spines, or spicules, but with scales or granules 14
6a Girdle much wider at the anterior end than elsewhere around the body; restricted to the open coast *Placiphorella velata* Dall, 1879
6b Girdle not much wider at the anterior end than elsewhere around the body; includes a number of species found in inland waters or protected bays (*Nuttallina californica*, however, is restricted to the open coast) 7
7a Girdle with hairs or flexible bristles, but without stiff spines or spicules (unless they occur in conjunction with hairs or bristles) 9
7b Girdle without hairs or flexible bristles, but with either stiff spines or microsopic, glassy bristles 8
8a Girdle with close-set brown and white spines; valves nearly as long as wide; sculpture of valves consisting of well-developed ribs that become more widely spaced as they approach the anterior margins of the valves (however, the valves may be badly eroded); length up to about 5 cm, and generally more than twice the width *Nuttallina californica* (Reeve, 1847)
8b Girdle with microscopic, glassy spicules (these are scattered); valves not nearly as long as wide; sculpture of valves consisting of rows of rounded tubercles (the tubercles are usually smaller and more numerous near the center of the valves than in lateral areas); length up to about 2 cm, and not more than twice the width *Chaetopleura gemma* Dall, 1879
9a With some bristles in the narrow interspaces between valves, and in a row bordering each of the end valves 10
9b Bristles of the girdle limited to lateral areas, not present in the

narrow interspaces between valves, and not usually forming a distinct row bordering each of the end valves 11

10a Valves (unless eroded) with some ribs consisting of coarse tubercles; color of girdle and bristles dark gray or blackish *Mopalia imporcata* Carpenter, 1864

10b Valves with only fine granules; color of girdle primarily green or some shade of red, or a mixture of green and red tones *Basiliochiton heathii* (Pilsbry, 1898)

11a Girdle bristles thick and stiff *Mopalia muscosa* (Gould, 1846)

11b Girdle bristles hairlike and flexible 12

12a Girdle with a distinct cleft at the posterior end; bristles usually not arising from light spots on the girdle 13

12b Girdle without a distinct cleft at the posterior end; bristles usually arising from light spots on the girdle *Mopalia lignosa* (Gould, 1846)

13a Girdle bristles somewhat flattened, with sharp-tipped white branches, and also with small spicules at their bases; interior of valves usually bluish green (sometimes tinged with lilac) or whitish tinged with rose *Mopalia ciliata* (Sowerby, 1840)

13b Girdle bristles neither flattened nor branched, and without spicules at their bases; interior of valves usually white *Mopalia hindsii* (Reeve, 1847)

14a Girdle with overlapping scales 15

14b Girdle with fine granulations, appearing velvety under magnification 19

15a Valves with coarse ribs (there are up to 9 of these on the end valves, and 1 or 2 on each side of valves 2 to 7) (prevailing external coloration of valves generally greenish or brownish; length up to about 4 cm, and usually slightly more than twice the width) *Callistochiton crassicostatus* Pilsbry, 1893

15b Valves without coarse ribs (but delicate ribs are usually present, those on the end valves being rather numerous) 16

16a Length up to about 6 cm, and decidedly more than twice the width (prevailing external coloration of valves reddish, rose, or buff) *Stenoplax fallax* (Pilsbry, 1892)

16b Length rarely exceeding 4 cm and not more than twice the width 17

17a Central areas of valves characterized by the presence of numerous pits (these may not be visible without magnification) and by the absence of rows of beadlike elevations 18

17b Central areas of valves characterized by the absence of pits and by the presence of rows of beadlike elevations 19

18a Pits of central portions of valves squarish, forming an almost networklike pattern *Ischnochiton retiporosus* Carpenter, 1864

18b Pits of central portions of valves not squarish and not forming an almost networklike pattern *Ischnochiton interstinctus* (Gould, 1846)

19a Prevailing coloration of dorsal surface reddish, or reddish and tan *Lepidozona mertensii* (Middendorff, 1846)

19b Prevailing coloration of dorsal surface greenish gray *Lepidozona cooperi* (Carpenter in Pilsbry, 1892)

20a Granulations of girdle rather coarse and somewhat like sand grains; insertion plates of the valves (the thin extensions of the valves that are buried in the tissue) without slits; sculpture of valves consisting mostly of granules or beads arranged in rows *Leptochiton internexus* Carpenter in Pilsbry, 1892

20b Granulations of girdle very fine, collectively presenting an almost velvety appearance; insertion plates of valves with slits; sculpture of valves consisting of protuberances that are not arranged in distinct rows 21

21a Lateral areas of valves with microscopic granulations, but without wartlike protuberances; length up to about 2 cm; intertidal and subtidal *Cyanoplax dentiens* (Gould, 1846)

21b Lateral areas of valves with microscopic wartlike protuberances; length up to about 3 cm; strictly subtidal *Cyanoplax hartwegii* (Carpenter, 1855)

Berry, S. S. 1917. Notes on west American chitons--I. Proc. Calif. Acad. Sci., ser. 4, 7:229-48.

------. 1919. Notes on west American chitons--II. Proc. Calif. Acad. Sci., ser. 4, 9:1-36.

------. 1927. Some notes on British Columbian chitons. Proc. Malac. Soc. London, 17:159-64.

------. 1951. Notes on some British Columbian chitons--II. Proc. Malac. Soc. London, 28:213-29.

Burghardt, G. E., & L. E. Burghardt. 1969. A Collector's Guide to West Coast Chitons. Special Publ. 4, San Francisco Aquarium Society. 45 pp.

Dall, W. H. 1899. Report on the limpets and chitons of the Alaskan and Arctic regions, with descriptions of genera and species believed to be new. Proc. U. S. Nat. Mus., 1:281-344.

Pilsbry, H. A. 1892-94. Polyplacophora (Chitons). *In* Manual of Conchology, 14:i-xxxiv, 1-350; 15:1-133.

Smith, A. G., & I. McT. Cowan. 1966. A new deep-water chiton from the northeastern Pacific. Occ. Pap. Calif. Acad. Sci., no. 56. 15 pp.

CLASS GASTROPODA

SUBCLASS PROSOBRANCHIA

Although most gastropods with shells belong in this group, a few opisthobranchs and pulmonates found in our area also have external shells. These are included in the key because their systematic position will not be apparent if only the shell has been conserved.

Some unusual species of prosobranchs that are parasitic in sea cucumbers and that have no shells are discussed following the key to shelled gastropods.

Key to Gastropods with Shells

1a Shell permanently cemented to the substrate, coiled loosely or irregularly, and usually twisted, resembling the tube of a worm more than a snail shell 2

1b Shell not permanently cemented to the substrate, either tubular, conical, cap-shaped, or tightly coiled 4

2a Shell loosely coiled into a flat spiral; generally with a prominent external ridge *Dendropoma lituella* (Mörch, 1861)

2b Shell either wound into an irregular spiral, or forming a contorted mass in which coiling is indistinct 3

3a Shell generally wound into an irregular spiral; shell with 2 internal shelflike ridges following the columella

Petaloconchus ?montereyensis Dall, 1919

3b Shell forming a contorted mass in which coiling is indistinct; shell without internal ridges *Vermetus compactus* (Carpenter, 1864)

4a Shell tubular, conical, or cap-shaped 5

4b Shell spirally coiled 38

5a Shell tubular, slightly curved 6

5b Shell conical or cap-shaped (sometimes irregular) 7

6a Shell smooth, brown *Fartulum occidentale* Bartsch, 1920

6b Shell with numerous contiguous rings, whitish *Caecum crebricinctum* Carpenter, 1864

7a Shell with a dorsal opening at or near the apex 8

7b Shell without a dorsal opening 12

8a Dorsal opening in the shell a slit, just anterior to the apex 9

8b Dorsal opening in the shell broadly oval, at the apex 11

9a Radial ribs of nearly uniform size; apex in the middle third of the shell; margin of shell smooth *Puncturella galeata* (Gould, 1846)

9b Radial ribs of 2 sizes; apex in the anterior third of the shell; margin of shell uneven or crenulate 10

10a With 14 to 23 coarse ribs, separated by much smaller ribs (these may be almost obscure); margin of shell crenulate *Puncturella cucullata* (Gould, 1846)

10b With more than 23 coarse ribs, separated by slightly smaller ribs; margin of shell uneven but not distinctly crenulate *Puncturella multistriata* Dall, 1914

Other species of this group have been reported for our area: *P. major* Dall, 1891; *P. cooperi* Carpenter, 1864; *P. decorata* Cowan & McLean, 1968; *Scelidotoma bella* (Gabb, 1865).

11a Length of opening about one-third the length of the shell; shell not often more than 2 cm long, too small to cover the animal *Megatebennus bimaculatus* (Dall, 1871)

11b Length of opening about one-tenth the length of the shell; shell often 5 cm long, large enough to cover the animal when it withdraws *Diodora aspera* (Rathke, 1833)

12a Interior of shell with a shelf or septum 13

12b Interior of shell without a shelf or septum 19

13a Shell conical, almost circular in outline; with a spiral septum, the free edge of which is twisted *Calyptraea fastigiata* Gould, 1846

13b Shell usually cap-shaped (sometimes nearly conical), generally not circular in outline and often slightly irregular; with a simple shelf 14

14a Shelf attached less on 1 side (the left side, as the aperture is turned toward the observer and the shelf is uppermost), so that it projects like a tongue *Crepipatella lingulata* (Gould, 1846)

14b Shelf attached to about the same extent on both right and left sides, its free margin nearly straight 15

15a Shell arched (frequently higher than wide), the apex prominent and beaklike; brown externally; on shells of *Tegula* and *Calliostoma* *Crepidula adunca* Sowerby, 1825

15b Shell low (never as high as wide), the apex, if distinct, not prominent; whitish or discolored, but not distinctly brown; frequently in shells, especially empty shells of bivalves, but not regularly associated with *Tegula* and *Calliostoma* 16

16a Apex of shell directed laterally and united with the margin (introduced with oysters from the Atlantic coast, and found only in a few

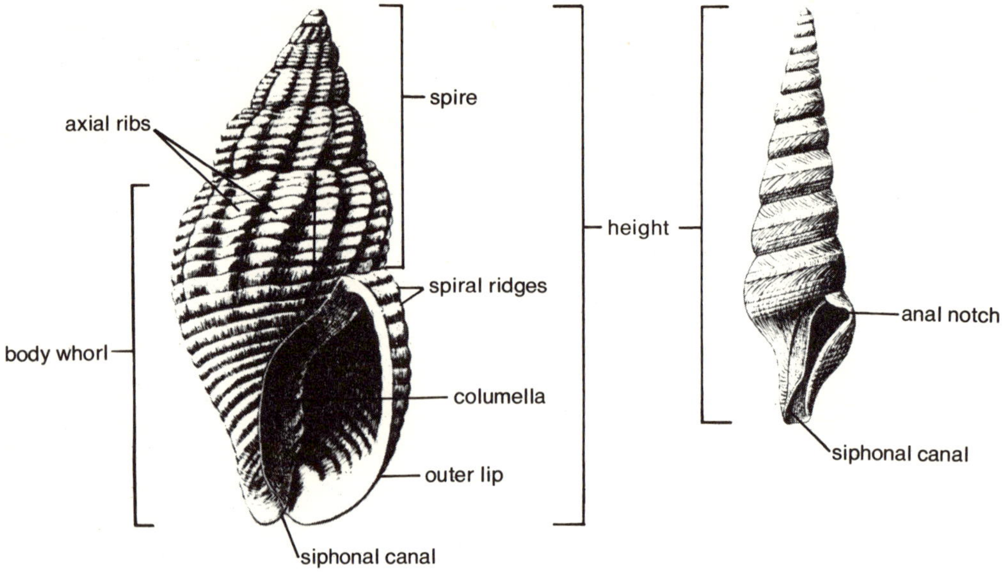

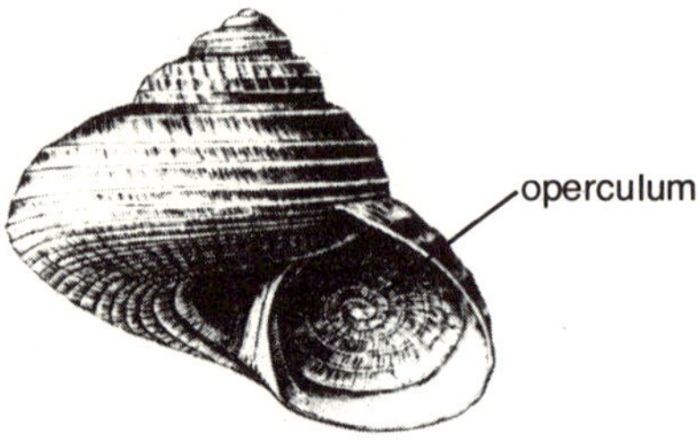

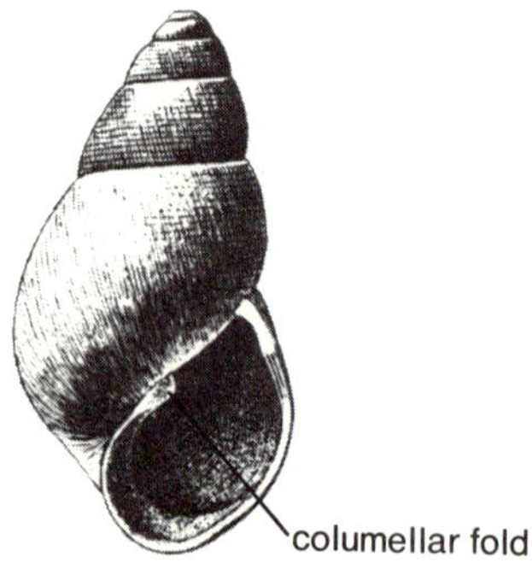

Odostomia
ORDER PYRAMIDELLACEA

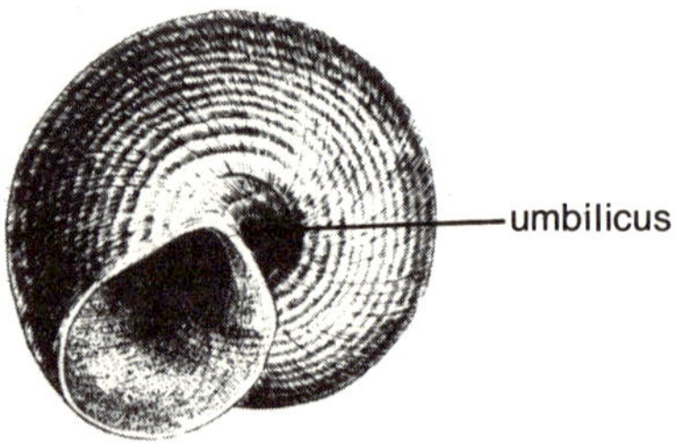

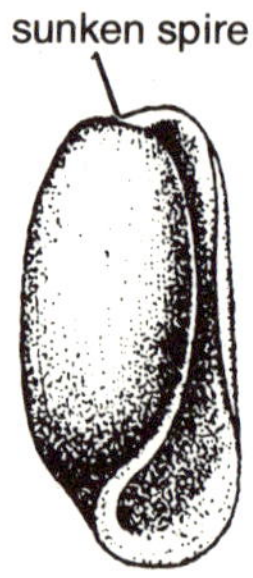

Cylichna
ORDER CEPHALASPIDEA

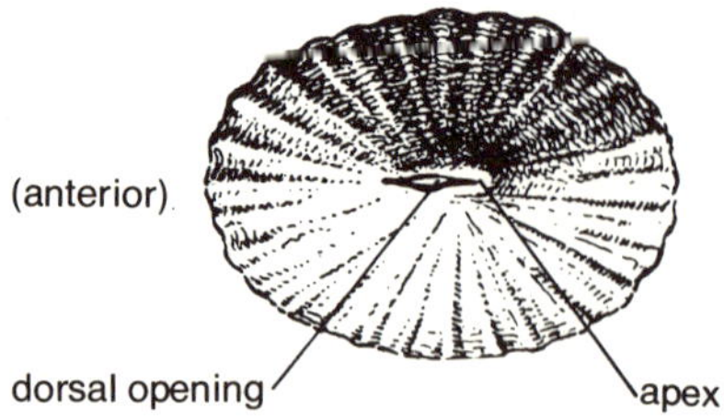

rather restricted areas) *Crepidula fornicata* Linnaeus, 1758
16b Apex of shell directed posteriorly, often overhanging the posterior margin of the shell 17
17a Shell with blunt spines *Crepidula aculeata* Gmelin, 1792
17b Shell without spines (but it may be slightly roughened) 18
18a Shell white, generally flat, with a shaggy yellow-brown periostracum; generally on rocks *Crepidula nummaria* Gould, 1846
18b Shell white, flat, concave, or misshapen (depending on habitat), with a thin brown periostracum (often in burrows of pholads, in empty shells, and in shells occupied by hermit crabs, fitting the contour of the concave surface) *Crepidula perforans* (Valenciennes, 1846)
19a Apex of the shell at or anterior to the middle (if only the empty shell is available, the open end of the horseshoe-shaped muscle scar will face anteriorly, except in 1 genus [*Siphonaria*] in which the muscle scar [often indistinct] is nearly c-shaped and opens on the animal's right side) 21
19b Apex of the shell posterior to the middle (nearer the end opposite that toward which the horseshoe-shaped muscle scar faces) 20
20a Apex very close to the posterior margin, often overhanging it; sculpture of external surface consisting of rather strong radial ribs, these intersecting faint concentric lines *Hipponix tumens* Carpenter, 1865
20b Apex generally about halfway between the middle and posterior margin; radial ribs weak, the concentric sculpture being more prominent *Hipponix cranioides* Carpenter, 1864
21a Outline of shell not symmetrical; apex slightly to the right of the midline; muscle scar (often indistinct) nearly c-shaped, opening on the animal's right side; with a shallow siphonal groove extending through the opening in the muscle scar *Siphonaria thersites* Carpenter, 1864 (Subclass Pulmonata)
21b Outline of shell symmetrical (unless deformed by injury); muscle scar horseshoe-shaped, its open end facing anteriorly; without a siphonal groove on the interior of the shell 22
22a Interior of shell whitish, without any color pattern; apex in the anterior third of the shell; exterior of shell with rather conspicuous concentric lines; subtidal *Cryptobranchia concentrica* (Middendorff, 1847)
22b Interior of shell usually with a color pattern of some sort (blotch in the apical region, marginal markings, etc.); apex in the anterior or middle third of the shell (in the middle third in *Acmaea mitra*, which is the only species with a shell that is white internally); concentric lines not conspicuous in most species; includes some species that are strictly intertidal, some that are subtidal, and some that are both intertidal and subtidal 23
23a Right and left margins of the shell (as it is viewed from above or below) nearly parallel, the outline of the aperture thus somewhat elongated 24
23b Right and left margins of the shell not nearly parallel, the outline of the aperture thus oval 28
24a Shell thin and fragile, light brown; apex close to the anterior margin; length up to about 8 mm, and about 4 times the width; restricted to leaves of *Phyllospadix* *Acmaea paleacea* (Gould, 1853)
24b Shell not especially thin and fragile and usually not light brown; apex nearer the middle than the anterior margin; length generally

greater than 8 mm (except in *A. triangularis*), and not more than twice the width; on algae, not regularly found on the leaves of *Phyllospadix* 25

25a Exterior of shell white, with radiating brown lines, or with a brown spot at the apex (sometimes on the interior as well as the exterior); length up to about 8 mm; on coralline algae or on snail shells encrusted by coralline algae (not likely to be found in our area) *Acmaea triangularis* (Carpenter, 1864)

25b Exterior of shell usually dark brown; length regularly greater than 1 cm, and often exceeding 2 cm; not on coralline algae or on shells encrusted by coralline algae 26

26a Right and left margins lower than the anterior and posterior margins, so the shell can be rocked when on a flat surface; on stipes of laminarian kelps, to which it clings after the fashion of a saddle *Collisella instabilis* (Gould, 1846)

26b Right and left margins at essentially the same horizontal level as the anterior and posterior margins, so the shell has little tendency to rock when on a flat surface; generally on the kelp *Egregia* 27

27a Interior of shell dark brown *Notoacmea insessa* (Hinds, 1842)

27b Interior of shell light blue *Collisella pelta* (Rathke, 1833) This is a form of *C. pelta* that mimics *N. insessa;* other forms of *C. pelta* are found in other habitats, and will be dealt with elsewhere in the key.

28a Length not exceeding 8 mm; exterior pink with yellow-brown dots, white streaks, and white dots; subtidal and generally associated with coralline algae; rare *Tectura rosacea* (Carpenter, 1864)

28b Length regularly greater than 1 cm; exterior not pink (unless overgrown by coralline algae); includes most of the more common intertidal species of limpets 29

29a Shell blackish brown, both internally and externally; length not often exceeding 1 cm; regularly on *Tegula funebralis* *Collisella asmi* (Middendorff, 1847)

29b Shell not blackish brown internally and externally; length generally exceeding 1 cm; not regularly on *Tegula funebralis* 30

30a Apex near the middle of the shell; shell heavy, white inside and out (unless overgrown by coralline algae, on which this species feeds); height of shell sometimes exceeding the width *Acmaea mitra* Rathke, 1833

30b Apex decidedly anterior to the middle of the shell; shell not especially heavy, and not white inside and out; height of shell not equal to the width 31

31a Height of shell not more than one-third the length (except perhaps in occasional individuals) 32

31b Height regularly more than one-third the length 34

32a Width of shell slightly more than three-fourths the length; shell not especially thin and fragile; length regularly exceeding 2 cm; exterior of shell rarely of a solid color, usually with light spots or streaks on a darker background 33

32b Width of shell about three-fourths the length; shell thin and fragile; length generally less than 1.5 cm; exterior sometimes of a solid color (white, yellow, brown, rose), sometimes with a pattern of oval white spots on a darker background (rare in our area) *Collisella ochracea* (Dall, 1871)

33a Height of shell generally about one-third the length, but may be slightly less; interior of shell, between the margin and the center, with a brown suffusion; exterior generally gray or olive, with white spots; rare in our area *Notoacmea fenestrata* (Reeve, 1855)

33b Height of shell decidedly less than one-third the length; interior of shell, between the margin and center, white to blue, not brown, exterior of shell generally gray or olive, with lighter spots or streaks; common *Notoacmea scutum* (Rathke, 1833)

34a Shell with coarse ridges radiating from the apex to the margins; margins distinctly scalloped 35

34b Shell without coarse ridges (if there are ridges, they are fine); margins not scalloped 36

35a Apex decidedly nearer the anterior end than the middle (frequently flush or even overhanging the anterior margin) *Collisella digitalis* (Rathke, 1833)

35b Apex about halfway between the anterior end and the middle *Collisella pelta* (Rathke, 1833)

36a Profile of shell, from apex to posterior margin, nearly straight 37

36b Profile of shell, from apex to posterior margin, markedly convex *Notoacmea persona* (Rathke, 1833)

37a Interior of shell bluish, with a dark border or series of dark spots along the margin, but generally without a brown blotch in the region of the apex; length rarely exceeding 1.5 cm *Collisella strigatella* (Carpenter, 1864)

37b Interior of shell bluish, with a dark border along the margin and also regularly with a brown blotch in the region of the apex; length commonly exceeding 3 cm *Collisella pelta* (Rathke, 1833)

38a Shell low and earlike, with a low spire near the posterior end, and with a series of holes (some closed) near the left side; length sometimes exceeding 10 cm *Haliotis kamtschatkana* Jonas, 1845

38b Shell generally not low and earlike, and without a series of holes near the left side; only a few species are larger than 10 cm, and all of these have tall spires 39

39a Outer lip of aperture with a deep slit; found only at depths greater than 500 m, and rare *Scissurella crispata* Fleming, 1832

39b Outer lip of aperture without a slit (however, there may be a notch or a canal at the anterior end of the aperture, or at both the anterior and posterior ends of the aperture); includes many intertidal and subtidal species 40

40a Length of aperture at least three-fourths the height of the shell, and sometimes equal to the height of the shell (in which case the spire is not raised above the level of the body whorl; it may even be sunken 41

40b Length of aperture less than three-fourths the height of the shell 42

41a Shell generally thin (sometimes so thin that it will break if pinched lightly); without teeth on the inside of the outer lip of the aperture; length of aperture either as long as the shell is high (in which case the spire is not raised) or about three-fourths the height of the shell (in which case there is a short spire); aperture usually decidedly wider in its lower half than elsewhere; height generally exceeding 3 mm Subclass Opisthobranchia, Order Cephalaspidea

41b Shell thick, similar in shape to a cowrie; with a series of teeth on the inside of the outer lip of the aperture; length of aperture nearly as long as the shell is high (spire indistinct); aperture not decidedly wider in its lower half than elsewhere; height about 2 mm

Granulina margaritula (Carpenter, 1857)

42a Periostracum elaborated into conspicuous hairs or bristles 43

42b Periostracum not elaborated into conspicuous hairs or bristles 44

43a Siphonal canal about one-third the total length of the aperture; height up to about 12 cm *Fusitriton oregonensis* (Redfield, 1848)

43b Without a distinct siphonal canal, but the anterior part of the aperture is angled in such a way that it forms a small spout; height seldom exceeding 3 cm *Trichotropis cancellata* Hinds, 1843

44a Anterior end of aperture with a distinct notch, short spout (as on a cream pitcher), or canal through which a siphon is extended 91

44b Anterior end of aperture without a notch, spout, or canal 45

45a Inside of shell pearly; with an operculum 46

45b Inside of shell porcelaneous rather than pearly, although it may be colored; usually with an operculum, but this is absent in certain genera 62

46a Columella with 1 or 2 small nodes (the nodes are weakly developed in species of *Homalopoma*, which have a calcareous operculum; the only other species having both a pearly shell interior and calcareous operculum is the large and distinctive *Astraea gibberosa* [choice 50a]) 47

46b Columella without any nodes 50

47a Operculum thin and horny, with numerous spiral lines; nodes on the columella rather prominent 48

47b Operculum calcified and rather thick in proportion to the relatively small shell, with only a few spiral lines; nodes on the columella weakly developed 49

48a Umbilicus closed by a callus; periostracum purple to nearly black; restricted to the outer coast, where it is often abundant at upper levels of rocky intertidal areas *Tegula funebralis* (A. Adams, 1855)

48b Umbilicus decidedly open; periostracum brown, usually with some lighter blotches and orange bands on the body whorl; mostly on the open coast, where it occurs at lower levels of the intertidal zone, but occasionally taken in somewhat protected situations

Tegula pulligo (Gmelin, 1791)

49a Shell nearly smooth, with weak spiral lines

Homalopoma baculum (Carpenter, 1864)

49b Shell with rather prominent spiral ridges (about 15 on the body whorl)

Homalopoma luridum (Dall, 1885)

Other species of *Homalopoma* perhaps also occur in our area.

50a Whorls bordered by a continuous wavy excrescence, at the base of which are 2 rows of beadlike protuberances; whorls with prominent radial ribs; operculum heavily calcified; diameter sometimes exceeding 5 cm; restricted to the open coast *Astraea gibberosa* (Dillwyn, 1817)

50b Whorls not bordered by a continuous wavy excrescence; whorls marked by spiral lines, without radial ribs; operculum thin and horny; diameter rarely exceeding 2.5 cm, and in most species no greater than 2 cm; includes some species found in inland waters 51

51a Umbilicus decidedly open 52

51b Umbilicus closed 57

52a Umbilical area bordered by a beaded or crenulate spiral ridge that is decidedly more conspicuous than the other spiral ridges of the basal portion of the body whorl; subtidal 53

52b Umbilical area not bordered by a beaded or crenulate spiral ridge that is decidedly more conspicuous than the other spiral ridges of the basal portion of the body whorls; includes subtidal and intertidal species 54

53a With 5 whorls; color faintly reddish brown *Solariella obscura* (Couthouy, 1838)

53b With 6 or 7 whorls; color generally tan, with light markings of a mauve or purplish brown color *Solariella peramabilis* Carpenter, 1864
Other species of *Solariella* are reported from our area, but they are rare and are not likely to be encountered except in deep water.

54a With distinct, closely-spaced spiral ridges; whorls convex, but not markedly so 55

54b Without spiral ridges; whorls markedly convex *Margarites helicinus* (Phipps, 1774)

55a Basal portion of body whorl almost flat; height usually about equal to the width; height up to about 1.5 cm in the largest species 56

55b Basal portion of body whorl rather distinctly convex; height usually a little greater than the width; height up to about 7 mm *Lirularia lirulata* (Carpenter, 1864)

56a Height frequently up to 1.5 cm (rarely exceeding this); spiral ridges generally rather prominent; exterior generally pinkish or purplish *Margarites pupillus* (Gould, 1849)

56b Height not over 6 or 7 mm; spiral ridges sometimes prominent, sometimes weak; exterior usually grayish *Lirularia succincta* (Carpenter, 1864)

57a Shell without spiral ridges, and with an almost waxy appearance (color usually white, yellowish, or pinkish); not often encountered at depths shallower than 100 m, and extremely rare *Calliostoma platinum* Dall, 1890

57b Shell with spiral ridges, and without a waxy appearance; includes some species found intertidally or subtidally in shallow water 58

58a Spiral ridges distinctly beaded 59

58b Spiral ridges not beaded 61

59a Beads on spiral ridges pronounced; base of body whorl decidedly convex; sutures between whorls decidedly indented; periostracum brown or gray (in deep water) *Cidarina cidaris* (A. Adams, 1864)

59b Beads on spiral ridges small and delicate; base of body whorl nearly flat; sutures between whorls scarcely indented; periostracum not typically brown or gray 60

60a Shell typically orange-yellow, dotted with brown, and with a bright purple or violet band encircling the lower edge of each whorl; with 8 or 9 whorls; body of living animal orange, with brown dorsal spots *Calliostoma annulatum* (Lightfoot, 1786)

60b Shell almost uniformly yellowish, tan, or pale pink; with 6 to 8 whorls; body of living animal cream, with brown dorsal spots *Calliostoma variegatum* Carpenter, 1864

61a Whorls, as the shell is viewed in profile, almost flat; periostracum more nearly tan or brown than pinkish brown or reddish brown; not common *Calliostoma canaliculatum* (Lightfoot, 1786)

61b Whorls slightly convex, periostracum generally pinkish brown or reddish brown; extremely abundant *Calliostoma ligatum* (Gould, 1849)

62a Height of shell no greater than the diameter 63

62b Height of shell greater than the diameter 66

63a Shell bluish white; umbilicus wide; with an operculum; aperture almost circular, equal to about half the total diameter; total diameter up to about 3 mm *Vitrinella columbiana* (Bartsch, 1921)

63b Shell either whitish and translucent, or pink and covered by an almost velvety periostracum; without an umbilicus; without an operculum; aperture decidedly more than half the total diameter; total diameter may exceed 1 cm 64

64a Shell whitish and translucent, without a periostracum; on compound ascidians, especially *Didemnum* *Lamellaria stearnsii* Dall, 1871
L. rhombica Dall, 1871 is sometimes found on solitary ascidians (*Ascidia paratropa, Pyura haustor*).

64b Shell pink, covered with a yellow or brown, almost velvety periostracum; not associated with compound ascidians 65

65a Periostracum yellow; greatest diameter up to about 1 cm *Velutina prolongata* Carpenter, 1865

65b Periostracum brown; greatest diameter up to about 1.5 cm *Velutina laevigata* (Linnaeus, 1767)

66a Columella with 1 or more distinct folds or ridges; height under 1 cm 67

66b Columella without any distinct folds; height may exceed 1 cm 69

67a Columella with 3 folds (one of which is weakly developed); shell without spiral ridges; generally chestnut-brown or yellowish brown, except where eroded (juveniles often have small hairs on the spire); abundant in salt marshes, at levels rarely inundated, among *Salicornia* and under debris *Phytia myosotis* (Draparnaud, 1801) (Subclass Pulmonata)

67b Columella with 1 to 3 folds; shell with or without spiral ridges; generally white or almost white but certain species may be light brown or gray; not found in salt marshes 68

68a Furrows between spiral ridges pitted; aperture about two-thirds the height of the shell, with 2 dark brown or blackish bands, each the width of about 6 spiral ridges, on the body whorls, and similar bands on the spire; with a single fold on the columella; in eelgrass flats *Rictaxis punctocoelatus* (Carpenter, 1864) (Subclass Opisthobranchia, Order Cephalaspidea)

68b Furrows between spiral ridges (if there is any spiral sculpturing) not pitted; aperture not more than half the height of the shell; without broad, dark brown or blackish spiral bands on the body whorl or other whorls; with 1 to 3 folds on the columella; not likely to be found in eelgrass flats Subclass Opisthobranchia, Order Pyramidellacea

69a Outline of shell, as viewed from one side, ranging from nearly globose to about twice as high as wide (higher than this in *Alvinia* and *Barleeia*); spire with not more than 6 distinct whorls 70

69b Outline of shell, as viewed from one side, usually more than twice as long as wide; spire usually with more than 6 whorls 83

70a Height of shell not exceeding 4 mm; generally either in salt marshes, or in sand and gravel 71

70b Height of shell regularly exceeding 5 mm; generally on rocky substrates or on algae and other vegetation, but rarely in sand and gravel, and, except for *Littorina*, not likely to be found in salt marshes 74

71a Inner lip of aperture forming a rather broad callus; shell less than twice as high as wide; dark brown; under debris and among vegetation at higher levels in salt marshes or at the borders of lagoons *Assiminea californica* (Tryon, 1865)

71b Inner lip of the aperture not forming a broad callus; shell more than twice as high as wide; whitish or brown; generally in sand and gravel, not likely to be found in salt marshes 72

72a Shell smooth; aperture more nearly oval than circular; operculum calcareous 73

72b Shell with strong spiral ridges (and sometimes also with axial ribs, thus cancellate); aperture nearly circular; operculum horny *Alvinia* spp.

73a Whorls almost flat; shell mostly light and shiny, with darker axial markings *Barleeia acuta* (Carpenter, 1864)

73b Whorls obviously convex; shell dark brown *Barleeia haliotiphila* Carpenter, 1864

74a Height of shell regularly more than 2 cm; umbilicus either a conspicuous opening or closed over by an obvious callus pad 75

74b Height of shell not exceeding 2 cm; umbilicus either a narrow slit between the aperture and the body whorl or absent (but not closed over by an obvious callus pad) 79

75a With a decidedly open umbilicus; operculum horny 76

75b Umbilicus closed by a callus; operculum calcified 78

76a Height of shell up to slightly more than 3 cm; with 4 or 5 whorls; color usually white or cream *Polinices pallidus* Broderip & Sowerby, 1829

76b Height of shell regularly greater than 5 cm; with 5 or 6 whorls; color cream to beige or pinkish 77

77a Operculum about 1 and one-half times as high as wide, its upper quarter almost triangular; umbilicus rather narrow, not extending nearly to the apex of the shell; height frequently attaining 10 cm; abundant intertidally on sand flats, and also subtidally *Polinices lewisii* (Gould, 1847)

77b Operculum about twice as high as wide, its upper quarter not nearly triangular; umbilicus broad, and extending nearly to the apex of the shell; height up to about 5 cm; subtidal and not likely to be found except near the northern limits of our area *Polinices draconis* Dall, 1903

78a Height of shell up to 4 cm; with not more than 5 whorls; found in shallow subtidal situations *Natica clausa* Broderip & Sowerby, 1829

78b Height of shell regularly exceeding 4 cm; with 6 or more whorls; found only in deep water *Natica russa* Gould, 1859

79a Umbilicus present in the form of a slit between the columella and the body whorl; generally in eelgrass or kelps at moderate and lower levels of the intertidal zone, and also subtidally 80

79b Umbilicus absent; typically on rocks at higher levels of the intertidal zone, but also found at higher tide levels in salt marshes 81

80a Shell (adult) higher than wide; spire (as the shell is viewed with the aperture facing the observer) about one-fourth the total height of the shell; height up to about 6 mm; generally on eelgrass *Lacuna variegata* Carpenter, 1864

80b Shell (adult) not quite as high as wide; spire about one-fifth the total height of the shell; height up to about 1 cm; generally on kelps *Lacuna carinata* Gould, 1849

81a Shell almost as wide as high, often with rather strong spiral sculpture; height up to almost 2 cm (common, often in association with *Littorina scutulata* [except in salt marshes, where it is generally the only species of *Littorina*]) *Littorina sitkana* Philippi, 1845

81b Shell decidedly higher than wide, without obvious spiral sculpture; maximum height generally not more than 1.3 cm 82

82a Spire generally consisting of 3 whorls; with a white band inside the aperture; color (except where eroded) usually brown or nearly olive, with irregular lighter mottling; rare in our area (a southern species) *Littorina planaxis* Philippi, 1847

82b Spire generally consisting of 4 whorls; without a white band inside the aperture; color (except where eroded) usually dark brown, purple, or black, often with a checkerboard pattern of whitish areas; common *Littorina scutulata* Gould, 1849

83a Shell toothlike, highly polished, without sculpturing other than growth lines; sutures between whorls so slightly indented that they are barely evident (moreover, the whorls themselves are almost flat); color white or faintly bluish white 84

83b Shell not toothlike, and not highly polished, usually with spiral or axial sculpturing; sutures between whorls indented and distinct; color generally not white or bluish white 86

84a Shell straight, up to about 1.3 cm high *Balcis micans* (Carpenter, 1864)

84b Shell decidedly bent, not more than 1 cm high 85

85a Shell about 3 and one-half to 4 times as long as wide, bent twice *Balcis columbiana* (Bartsch, 1917)

85b Shell about 2 and one-half times as long as wide, bent only once *Balcis montereyensis* (Bartsch, 1917)

Other species of *Balcis*, especially *B. randolphi* (Vanatta, 1899) (similar to *B. montereyensis*) and *B. rutila* (Carpenter, 1864) (similar to *B. micans*) may be expected.

86a With both axial ribs and spiral ridges, the intersections forming rather strong beads *Tachyrhynchus lacteolus* (Carpenter, 1864)

86b Sculpture limited to axial ribs (except for a single angular ridge near the base of the body whorl in *Opalia*) 87

87a As the shell is viewed with the aperture lowermost and facing the observer, the axial ribs run more nearly parallel to the left side of the shell than to the right side; aperture nearly semicircular *Rissoina newcombei* Dall, 1897

87b As the shell is viewed with the aperture lowermost and facing the observer, the axial ribs run more nearly parallel to the right side of the shell than to the left side; aperture almost circular 88

88a With a spiral ridge near the base of the body whorl (the axial ribs may terminate at this keel) 89

88b Without a spiral ridge near the base of the body whorl (the axial ribs extend quite to the base of the body whorl) 90

89a Spiral ridge on the body whorl so well developed that it projects; axial ribs on body whorl crossing the ridge and thus extending to the lowermost portion of the body whorl *Opalia montereyensis* (Dall, 1907)

89b Spiral ridge on the body whorl not so well developed that it projects; axial ribs on the body whorl terminating at the ridge *Opalia chacei* Strong, 1937

90a With 10 whorls (unless 1 or 2 have been eroded, which is usually the case); shell almost completely white (in any case, no purplish tones are present); height occasionally exceeding 2 cm *Epitonium indianorum* (Carpenter, 1864)

90b With not more than 8 whorls (and with fewer if any have been eroded); shell generally with a purplish band below the sutures between whorls; height not often attaining 1.5 cm *Epitonium tinctum* (Carpenter, 1864)

*91a Aperture with a notch or trace of a notch on the upper portion of the outer lip. (This is called the anal notch or posterior notch, and is characteristic of members of the family Turridae. Even if a notch is not obvious, the lines of axial sculpture will usually indicate that an indentation was present earlier.) 92

91b Aperture without a notch or trace of a notch on the upper portion of the outer lip 101

92a Shell with pronounced axial ribs 93

92b Shell without pronounced axial ribs (but axial lines may be present) 100

93a Exterior of shell gray, brown, violet, pinkish violet, or maroon (definitely not white, or white with a pink body whorl, and without spiral bands of color) *Oenopota levidensis* (Carpenter, 1864)

93b Exterior of shell white, or white with a pink body whorl, or white with spiral bands of a darker color (generally gray, brownish gray, or violet) 94

94a Shell with brown, brownish gray, or violet spiral bands *Kurtziella plumbea* (Hinds, 1843)

94b Shell without brown, brownish gray, or violet spiral bands 95

95a Height of shell about twice the width, and less than twice the height of the aperture *Oenopota excurvata* (Carpenter, 1864)

95b Height of shell about 2 and one-half to 3 times the width and more than twice the height of the aperture 96

96a Axial ribs decidedly more prominent than the spiral lines, and with strongly raised nodes at the shoulders of the whorls *Oenopota turricula* (Montagu, 1803)

96b Axial ribs not decidedly more prominent than the spiral lines (however, they may be a little more prominent), and with only slight thickenings at the shoulders of the whorls 97

97a Shell decidedly tabulate (upper portion of each whorl with a flattened area that meets the base of the next whorl at an angle of about 90°) *Oenopota tabulata* (Carpenter, 1865)

97b Shell not tabulate (however, the upper portion of each whorl may have a flattened shoulder that meets the base of the next whorl at an angle of about 45°) 98

98a Axial ribs nearly straight *Oenopota fidicula* (Gould, 1849)

98b Axial ribs sinuous, especially on the body whorl 99

99a Upper portions of whorls with distinctly flattened shoulders *Oenopota excurvata* (Carpenter, 1864)

99b Upper portions of whorls without distinctly flattened shoulders *Oenopota pleurotomaria* (Couthouy, 1839)

100a Shell white (but it may have some black periostracum near the base); with sufficient axial sculpturing to interact with the spiral lines to present a cancellate appearance *Ophiodermella cancellata* (Carpenter, 1864)

100b Shell brown, or with alternating brown and white spiral bands; axial lines weak, the surface of the shell thus not noticeably cancellate 101

101a Inside of aperture white (shell may be sinistral or dextral!); maximum size about 5 cm; at depths greater than 30 m *Antiplanes perversa* (Gabb, 1865)

101b Inside of aperture violet or brown; maximum size about 3.6 cm; low

*Couplets 91 to 101, dealing with turrids, were prepared by Ronald Shimek.

intertidal and shallow subtidal
Ophiodermella incisa (Carpenter, 1864)

102a Shell highly polished, without any sculpturing other than fine growth lines; widest part of aperture (near its anterior end) less than half as wide as the shell; operculum inconspicuous; typically on sandy beaches and subtidally on sandy bottoms 103

102b Shell not highly polished, and usually with at least some sculpturing; widest part of aperture (generally near its middle) at least half as wide as the shell; operculum usually conspicuous and large enough to close the aperture; not necessarily on sandy substrates (although some species are typically found on such substrates) 104

103a Shell about twice as high as wide; background color variable, but usually a warm gray or purplish gray, with brown or purplish lines at the sutures; restricted to sandy beaches of the open coast
Olivella biplicata (Sowerby, 1825)

103b Shell about 2 and one-half times as high as wide; background color usually cream or pale tan, with some reddish brown markings and brown or purple lines at the sutures; on some protected sandy beaches as well as the open coast, and also subtidal
Olivella baetica Carpenter, 1864

104a Lowermost portion of body whorl, including the siphonal canal, set off from the rest of it by a conspicuous groove 105

104b Lowermost portion of body whorl not set off from the rest of it by a conspicuous groove 107

105a Inner lip of the aperture with an orange callus that spreads out over the body whorl; axial ribs on body whorl not extending to the groove at the base *Nassarius fossatus* (Gould, 1850)

105b Inner lip of the aperture without a large orange callus; axial ribs on the body whorl reaching to the groove at the base 106

106a Upper portions of whorls, next to the sutures, with narrow, flattened shoulders; axial and spiral lines about equal, the sculpture thus appearing cancellate *Nassarius perpinguis* (Hinds, 1844)

106b Upper portions of whorls without flattened shoulders; axial ribs widely spaced and more pronounced than the spiral lines, the sculpture thus not appearing cancellate *Nassarius mendicus* (Gould, 1850)

107a Edges of siphonal canal touching or nearly touching for at least part of the length of the canal, so that the latter is essentially a closed tube (some specimens of *Ocenebra interfossa* have an open canal [thus this species is covered twice, in choices 112a and 126b]; *O. lurida* [choice 111a] may also have an open canal) 108

107b Edges of siphonal canal not touching or even nearly touching, so that the canal is an open trough for its entire length 113

108a With a large tooth on the outer lip of the aperture; with 3 winglike axial excrescences, as well as unevenly spaced spiral ridges; height up to about 8 cm *Ceratostoma foliatum* (Gmelin, 1791)

108b Without a large tooth on the outer lip of the aperture (but there may be several small teeth on the lip); without winglike excrescences, but with coarse axial ribs and sometimes also with well developed spiral ridges; height not more than about 3 cm 109

109a Spiral ridges well developed on all whorls; outer lip of aperture never thickened to the extent that it is nearly as wide as the opening 110

109b Spiral sculpturing represented by faint lines; outer lip of aperture sometimes thickened to the extent that it is nearly as wide as the opening (locally introduced with the Japanese oyster, *Crassostrea gigas*) *Ocenebra japonica* (Dunker, 1860)

110a Height of shell decidedly more than twice the width *Trophonopsis lasius* (Dall, 1919)

110b Height of shell only about twice the width 111

111a Axial ribs prominent, but usually rather low and rounded, not sharply elevated (common intertidally) *Ocenebra lurida* (Middendorff, 1849)

111b Axial ribs usually narrow and sharply elevated 112

112a Spiral ridges generally narrower than the furrows between them; height not often exceeding 2 cm *Ocenebra interfossa* Carpenter, 1864

112b Spiral ridges at least as wide as the furrows between them; height up to about 4 cm *Ocenebra sclera* (Dall, 1919)

113a Shell with at least 8 distinct whorls, not counting 1 or 2 that may have been worn away at the apex (in specimens which have been abruptly worn down to about 4 or 5 whorls, one can estimate, on the basis of the shape of the surviving portion of the shell, that about 10 whorls have developed altogether) 114

113b Shell with no more than 7 whorls, even when none appears to have been worn away 117

114a Siphonal canal well developed, even though short; spiral ridges distinctly (and sometimes conspicuously) beaded where they are intersected by axial ribs 115

114b Siphonal canal poorly developed, often barely evident; spiral ridges not beaded in the most common species (*Bittium eschrichtii*), which lacks axial ribs (at least on the larger whorls), but beaded in some other species having axial ribs 116

115a Siphonal canal narrow, directed toward the left at a nearly right angle to the long axis of the shell (as the shell is viewed with the aperture lowermost and facing the observer); with about 12 axial ribs *Batillaria attramentaria* (Sowerby, 1855)

This has usually been called *B. zonalis* (Bruguière, 1792)

115b Siphonal canal broad, not markedly directed toward the left; with more than 18 (sometimes more than 30) axial ribs *Cerithiopsis*

Several species may be expected. A key to segregate them will be difficult to develop until it is clear which species are valid.

116a Without any axial ribs, at least on the larger whorls, the spiral ridges thus not beaded *Bittium eschrichtii* (Middendorff, 1849)

116b With axial ribs, the spiral ridges thus at least slightly beaded *Bittium* spp. (other than *B. eschrichtii*)

The following, among others, have been reported from our area: *B. attenuatum* Carpenter, 1864 (=*B. boreale* Bartsch, 1911); *B. munitum* (Carpenter, 1864); *B. vancouverense* Dall & Bartsch, 1910 (=*B. sanjuanensis* Bartsch, 1921; *B. challisiae* Bartsch, 1917). They are difficult to segregate in a key.

117a Shell with axial ribs (these may be low and inconspicuous, however) or thin axial lamellae, as well as spiral ridges 118

117b Shell without axial ribs or lamellae (sculpturing restricted to spiral ridges, and even these may be lacking) 130

118a Axial ribs limited to the spire 117

118b Axial ribs or lamellae extending to the body whorl (though they may be less distinct on the body whorl than on the spire) 120

119a Columella with a single fold (near the siphonal notch); periostracum blackish, thick, cracked, but otherwise more or less continuous except where the spire is eroded; inside of aperture purplish brown;

height up to about 2.5 cm; found intertidally in some bays where the Atlantic oyster, *Crassostrea virginica*, has been introduced *Ilyanassa obsoleta* (Say, 1822)
This is often called *Nassarius obsoletus*.

119b Columella without a fold; periostracum gray, essentially restricted to the furrows between spiral ridges; inside of aperture light, with the dark lines of the furrows showing through to the inside; height up to about 4 cm; common intertidally on rocky shores and on rocks in bays *Searlesia dira* (Reeve, 1846)

120a Axial ribs on the body whorl limited to the upper half of it 121

120b Axial ribs or lamellae on the body whorl present on at least much of the lower half of it 124

121a Outline of aperture nearly ellipsoidal; columella and inside of outer lip of the aperture with a series of little transverse folds; height up to about 3 cm; color light brown, yellowish brown, or pinkish brown; a common intertidal and shallow subtidal species *Amphissa columbiana* Dall, 1916

121b Outline of aperture (including the siphonal canal) more nearly pear-shaped than ellipsoidal; if the columella has folds, there are no more than 5 of these, and there are no folds on the inside of the outer lip of the aperture; height either up to about 2 cm or regularly exceeding 10 cm; color generally either brown (due to a nearly continuous periostracum) or nearly white on the spiral ridges, brown in the furrows; rather rare subtidal species 122

122a Height up to about 2 cm; axial ribs rather sharply etched; with 2 to 5 (generally 2 or 3) columellar folds *Admete couthouyi* Jay, 1839
A. modesta (Carpenter, 1865) also occurs in our area.

122b Height regularly exceeding 10 cm; axial ribs low and broad, almost like elongated tubercles; without folds on the columella (shell remarkably thin for such large gastropods) 123

123a Aperture, including the siphonal canal, about half the height of the shell; axial ridges prominent and not much wider than the furrows between them; periostracum thin, essentially limited to the furrows *Beringius undatus* (Dall, 1919)
B. crebricostatus Dall, 1897 lacks the axial ribs, and has fewer (?) spiral ridges ("14 or more" instead of "about 18") on the body whorl. *B. undatus* is perhaps just a variety of the latter, as Dall originally proposed.

123b Aperture more than half the height of the shell; spiral ridges not prominent (obscure or nearly so on the upper portions of the whorls), much wider than the furrows between them; periostracum fibrous, abundant, and continuous *Beringius eyerdami* Smith, 1959

124a Siphonal canal (measured from the angle in the outer lip of the aperture where the canal begins) one-third to one-half the total length of the aperture 125

124b Siphonal canal decidedly less than one-third the total length of the aperture 128

125a Axial ribs fairly broad, not at all like lamellae 126

125b Axial ribs rather thin, almost like lamellae (though not at all frilled) 127

126a Spiral ridges delicate, closely spaced; interior of aperture purple or brown; found in bays where the Atlantic oyster, *Crassostrea virginica*, has been introduced *Urosalpinx cinerea* Say, 1822

126b Spiral ridges rather strongly developed, widely spaced (the ridges are narrower than the furrows between them); interior of aperture whitish or yellowish; widely distributed on rocky shores *Ocenebra interfossa* Carpenter, 1864

127a Axial ribs on the body whorl and lower whorls of the spire with angular "shoulders" that generally project above the level of the suture with the next higher whorl *Boreotrophon orpheus* (Gould, 1829)

127b Axial ribs without angular "shoulders" *Boreotrophon pacificus* (Dall, 1902)

128a Axial sculpture consisting of thin, frilly lamellae; spiral sculpture (may be obscured by the lamellae) consisting of 1 or 2 prominent ridges on each whorl *Thais lamellosa* (Gmelin, 1791)
The frilled forms of *T. lamellosa* tend to be restricted to situations where the water is calm. Forms which lack the lamellae are regularly found in exposed situations (see choice 132b).

128b Axial sculpture consisting either of fairly broad ribs or what look like wrinkles (these may be discontinuous and irregular), but not of frilly lamellae; spiral sculpture generally consisting of closely spaced spiral ridges 129

129a Axial sculpture consisting of broad ribs; generally without folds on the columella; height various; includes intertidal and subtidal species 130

129b Axial sculpture consisting of what look like wrinkles (these are commonly discontinuous and irregular); with folds on the columella; height up to about 5 cm; subtidal *Buccinum plectrum* Stimpson, 1865

130a With 2 or more folds on the columella; restricted to deep water 131

130b Without folds in the columella; includes some intertidal and shallow subtidal species 132

131a With 2 distinct folds on the columella; without obvious spiral lines *Boreomelon stearnsii* (Dall, 1872)

131b With about 5 folds on the columella; with fine spiral lines *Volutomitra alaskana* Dall, 1902

132a Sculpture consisting of closely spaced spiral ridges (in *Mitrella*, these are faint and restricted to the lower half of the body whorl, close to the columella, and sometimes they are absent; in occasional specimens of *Thais emarginata*, which normally has spiral ridges over most of the shell, the ridges may be obscure); height rarely exceeding 3 cm 133

132b Sculpture consisting of only 1 or 2 prominent ridges on each whorl (the lower part of the body whorl generally has a number of smaller ridges), the whorls flattened and angled from the sutures to these ridges; height up to about 10 cm *Thais lamellosa* (Gmelin, 1791)

133a Height up to about 1 cm; spiral ridges faint and limited to the lower half of the body whorl (they are most likely to be visible close to the columella; to see them will require magnification) 134

133b Height regularly more than 1 cm; spiral ridges distributed over all of the larger whorls 136

134a Upper part of body whorl with a slight "shoulder," due to the presence of an angular keel; outer lip of aperture thickened and with a series of distinct teeth *Mitrella carinata* (Hinds, 1844)

134b Upper part of body whorl without a "shoulder;" outer lip of aperture not markedly thickened, with several very indistinct teeth 135

135a Whorls nearly flat; periostracum forming ridges running lengthwise on the shell *Mitrella tuberosa* (Carpenter, 1864)

135b Whorls convex, periostracum not forming ridges running lengthwise on the shell *Mitrella gouldi* (Carpenter, 1856)

136a Periostracum persistent, dark brown, continuous over the ridges and grooves; spiral ridges so low and so crowded that they are barely evident unless viewed with the aid of a magnifier (length up to about 5 cm; deep subtidal, rare) *Colus jordani* Dall, 1913
Other species of *Colus* will perhaps be encountered in deep water.

136b Spiral ridges generally prominent and separated by distinct furrows (except in some specimens of *Thais emarginata* [intertidal] which may have nearly smooth shells); periostracum not continuous (if present, largely restricted to the furrows between ridges), and not necessarily dark brown 137

137a Length under 4 cm; siphonal canal less than one-fourth the total length of the aperture; common intertidal species 138

137b Length regularly exceeding 8 cm; siphonal canal one fourth to one-third the total length of the aperture (the outer lip of the aperture is angled where the canal begins); strictly subtidal 139

138a Larger spiral ridges generally alternating almost regularly with smaller ones; whorls not usually separated from one another by a deep groove; color of periostracum (in the grooves between spiral ridges) generally brown or gray, but ranging from yellow to nearly black; height up to about 2 cm *Thais emarginata* (Deshayes, 1839)

138b Spiral ridges generally of nearly equal size; whorls separated from one another by a deep groove; periostracum (in the grooves between spiral ridges) usually gray; height up to or slightly exceeding 3 cm *Thais canaliculata* (Duclos, 1832)

139a Upper portions of whorls markedly flattened, forming terracelike shoulders; spiral ridges close together, larger ones alternating rather regularly with smaller ones; height up to about 11 cm *Neptunea tabulata* (Baird, 1863)

139b Upper portions of whorls not flattened into terracelike shoulders; spiral ridges, on the whole, widely spaced, mostly of about the same size (in any case, there is no regular alternation of larger and smaller ridges); height up to about 15 cm *Neptunea lirata* (Gmelin, 1791)

Parasitic Gastropods Without Shells

Family Thyonicolidae. These are wormlike parasites of sea cucumbers, most of which lie in the coelom but are attached to and covered by the intestinal peritoneum. Two species from this region were described as new by Tikasingh (1961). *Comenteroxenos parastichopoli* is found in *Parastichopus californicus*, and *Thyonicola americana* occurs in *Eupentacta quinquesemita* and *E. pseudoquinquesemita*. Kincaid (1964) questioned the validity of *C. parastichopoli*, indicating that it appears to be identical with *Enteroxenos oestergreni* Bonnevie, 1902, which parasitizes *Stichopus tremulus* in Europe. In a preliminary report Tikasingh (1960) stated that *Cucumaria miniata* and *C. lubrica* are parasitized by *Thyonicola mortenseni* Mandahl-Barth, 1941, not previously reported from North America, but he did not elaborate on this in his 1961 paper.

Demond, J. 1952. The Nassariidae of the west coast of North America between Cape San Lucas, Lower California, and Cape Flattery, Washington. Pacific Sci., 6:300-17.

Griffith, L. M. 1967. The Intertidal Univalves of British Columbia. Hand-

book 26, British Columbia Provincial Museum. 101 pp.
Kincaid, T. 1964. A gastropod parasitic in the holothurian, *Parastichopus californicus* (Stimpson). Trans. Amer. Micr. Soc., 83:373-76.
Tikasingh, E. S. 1960. Endoparasitic gastropods of some Puget Sound holothurians. J. Parasit., 46 (suppl.):13.
------. 1961. A new genus and two new species of endoparasitic gastropods from Puget Sound, Washington. J. Parasit., 47:268-72.

SUBCLASS OPISTHOBRANCHIA

Gordon A. Robilliard*
Woodward-Envicon, Inc., San Diego, California

1a With an external shell large enough to accommodate all (or nearly all) of the animal when it withdraws 2
1b Either without an external shell, or with a shell that is only partially visible externally and incapable of accommodating more than a small part of the animel 4
2a Shell sinistrally coiled, trochoid in shape Order Thecosomata
2b Shell either not coiled or coiled dextrally, not trochoid in shape 3

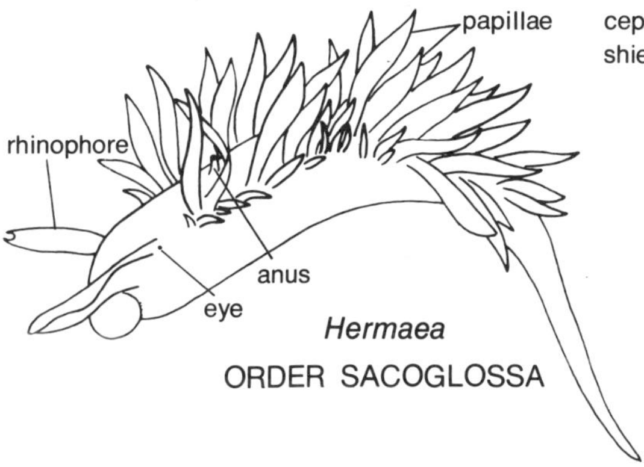

Hermaea
ORDER SACOGLOSSA

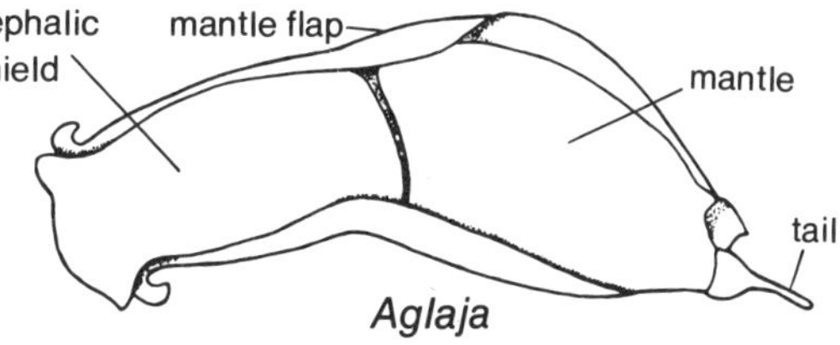

Aglaja
ORDER CEPHALASPIDEA

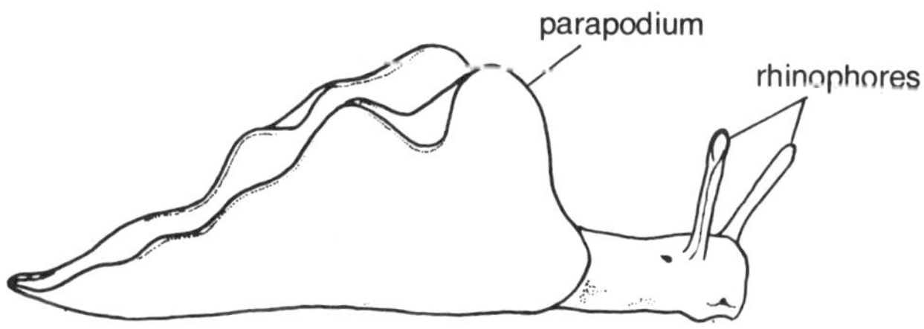

Elysia
ORDER SACOGLOSSA

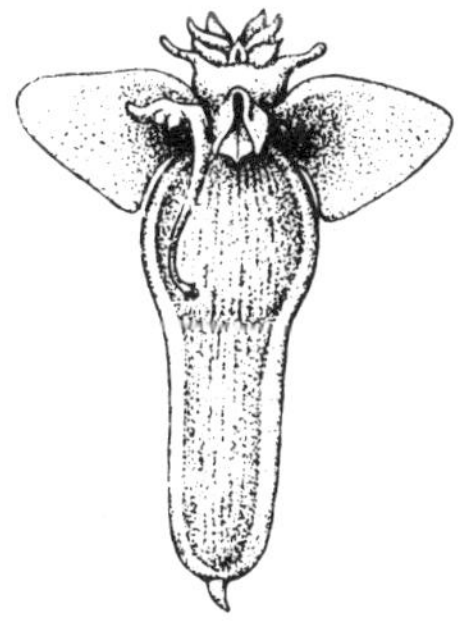
Clione
ORDER GYMNOSOMATA

3a Shell turriform or obovate; aperture less than half the total height of the shell Order Pyramidellacea
3b Shell cylindrical; aperture more than half the total height of the shell Order Cephalaspidea (in part)

*The author wishes to thank Sue Peterson (Woodward-Envicon Co., Inc.), Helen Robilliard, and Carol DeBoer for the help they gave in preparation of this section.

4a Permanently pelagic; parapodia flipperlike, highly specialized for swimming, and located just behind the head Order Gymnosomata
4b If pelagic, only temporarily so; parapodia, if present, not flipperlike, but forming sheets running most of the length of the animal and not highly specialized for swimming 5
5a With a single true ctenidium located on the right side of the body (between the overlapping mantle margin and a distinct foot) Order Notaspidea
5b Either without a true ctenidium, or with a ctenidium on both sides of the body (between the mantle margin and foot) 6
6a With a gill chamber (its length about one-fifth the total length of the body) located about halfway back on the dorsum, just to the right of the midline (the only species in our area has longitudinal black stripes on a green or yellow-green dorsum) Order Anaspidea
6b Without a gill chamber on the dorsum (none of the species in our area has black stripes on a green or yellow-green dorsum) 7
7a Shell present (although it may be small and delicate), partially or completely hidden by the mantle; cephalic shield present; fleshy parapodia reflected dorsally but overlapping one another only slightly, if at all Order Cephalaspidea (in part)
7b Shell absent; cephalic shield absent; parapodia, if present, wide and thin, and if reflected dorsally, overlapping one another considerably 8
8a Small (maximum length 3 to 4 mm), interstitial, in sand; visceral mass distinct from foot and head; visceral mass longer than the foot, and projecting posteriorly beyond it Order Acochlidiacea
8b Not especially small (regularly more than 3 to 4 mm long, except when young); not interstitial in sand; visceral mass not distinct from the foot and head, and not projecting beyond the posterior end of the foot 9
9a With a pneumostome, leading to a pulmonary sac, opening on the posteroventral side of the mantle immediately posterior to the anus; eyes at the ends of distinct stalks Order Onchidiacea
9b Without a pneumostome; eyes not at the ends of stalks 10
10a Rhinophores absent Order Sacoglossa (in part)
10b Rhinophores present 11
11a Clavus of the rhinophores perfoliate, or with longitudinal ridges, or with vertical pinnate plumes immediately beside the clavus Order Nudibranchia (in part)
11b Clavus of the rhinophores smooth and not distinct from the rhinophore stalk (in cross-section, the rhinophores may be solid or rolled into cylinders) 12
12a Rhinophores retractile into sheaths Order Nudibranchia (in part)
12b Rhinophores not retractile into sheaths 13
13a With elongate papillae on the dorsum 14
13b Without elongate papillae on the dorsum Order Sacoglossa (in part)
14a Rhinophores rolled into cylinders Order Sacoglossa (in part)
14b Rhinophores solid in cross-section 15
15a Anus in the midline, just posterior to the rhinophores Order Sacoglossa (in part)
15b Anus on the right side of the body Order Nudibranchia (in part)

ORDER NOTASPIDEA

Berthella californica (Dall, 1900) is the only species known from this area. It is white all over and strikingly visible on the rocky subtidal bottom. It apparently preys on ascidians, and perhaps also on sponges.

ORDER ANASPIDEA

Phyllaplysia taylori Dall, 1900 is the only anaspidean known from this area. It is found on eelgrass, *Zostera marina*, a background against which it is rather cryptically colored and patterned. (*Aplysia* and related genera ["sea hares"] have not been reported north of northern California.)

ORDER THECOSOMATA

Spiratella helicina (Phipps, 1774) (=*Limacina helicina*) is the most commonly observed thecosome in the region. Other species have been reported, but mostly on the basis of empty shells.

ORDER ACOCHLIDIACEA

One species, apparently undescribed, but tentatively placed in the genus *Hedylopsis*, has been found in sand collected at various localities in the San Juan Archipelago and in Puget Sound.

ORDER ONCHIDIACEA

Onchidella borealis Dall, 1871, a species with a difficult and confusing taxonomic past, is apparently the only onchidiacean in the region. It is restricted to higher levels of rocky intertidal areas, and is usually found in crevices, under rocks or algae, and in other places where it is able to remain damp despite long periods of exposure to the air.

ORDER PYRAMIDELLACEA

Numerous genera, subgenera, and species of northeast Pacific pyramidellids have been named, renamed, or placed in synonomy. The last major work on the group was that of Dall and Bartsch (1909), but considerable revision is probably in order. The key provided here covers the genera likely to be found in this area.

1a Aperture usually large, at least one-third the height of the shell; shell short, ovate or sub-conical, with only a few whorls (usually 4 to 6); shell usually sculptured, though many species are smooth *Odostomia*

1b Aperture usually small, less than one-fourth the height of the shell; shell elongate, conical or cylindrical; usually with more than 6 whorls; shall may or may not be sculptured 2

2a Shell conical; whorls usually inflated and regularly increasing in size toward the aperture; shell usually without sculpture; with 1 to 3 folds on the columella *Pyramidella*

2b Shell usually cylindrical or only slightly conical (however, a few species are distinctly conical); whorls not usually inflated, and increasing only slightly in size toward the aperture; shell usually sculptured; with only 1 fold on the columella *Turbonilla*

ORDER GYMNOSOMATA*

Only 2 gymnosomes have been reported from this area. Although there are marked anatomical differences between them, they may not be readily apparent because of the small size of the animals and because the distinctive buccal appendages are usually retracted. (If crystals of urethane are added to a small amount of water containing the specimens, the buccal appendages will usually be protruded.)

1a With stellate chromatophores on the body; lateral gill a membranous, unpigmented band near the middle of the body, posterior gill with 4 radial crests; with 2 lateral buccal appendages, each with numerous stalked suckers; proboscis large and protrusible *Crucibranchaea macrochira* (Meisenheimer, 1905)

1b Without chromatophores; without gills; with 3 pairs of buccal appendages (buccal cones), these without stalked suckers; without a proboscis *Clione limacina* (Phipps, 1773)

ORDER CEPHALASPIDEA

1a Shell large enough to accommodate the animal when it retracts 7

1b Shell either not visible externally, or much too small to accommodate the animal when it retracts 2

2a Shell visible on the dorsum, but partially covered by parapodia 3

2b Shell not visible externally 4

3a Shell with a small hole in the apex and with a relatively narrow aperture; body about half as long as the last whorl *Haminoea vesicula* Gould, 1855

3b Shell without a hole in the apex, and with a large, open aperture; body less than half as long as the last whorl *Haminoea virescens* (Sowerby, 1833)

4a Body completely dark purple or blue-black *Aglaja diomedea* (Bergh, 1894)

4b Body not completely dark purple or blue-black 5

5a Body light to dark reddish brown, with numerous yellow spots; with yellow lines on the edge of the cephalic shield, foot, and borders of the mantle flaps *Aglaja ocelligera* (Bergh, 1894)

5b Body not reddish brown, and without yellow spots; without yellow lines on the cephalic shield, foot, or borders of the mantle flap 6

6a Body pale yellow, with scattered clusters of small red dots *Gastropteron pacificum* Bergh, 1894

6b Body pale yellow to white, without clusters of small red dots *Philine* sp. (possibly *P. bakeri* Dall, 1919 or *P. polaris* Aurivillius, 1885)

7a Spire on shell sunken (inconspicuous) or concealed 8

7b Spire apparent, slightly or moderately elevated 9

8a Spire concealed by the apex *Volvulella* spp.

8b Spire sunken, but not concealed *Cylichna* spp.

9a Columella with a basal fold *Acteocina* spp. *A. culcitella* (Gould, 1853) and *A. eximia* (Baird, 1863) are probably

*Information for this section was contributed by Carol M. Lalli.

represented in our area.
9b Columella smooth *Retusa ?harpa* Dall, 1871

ORDER SACOGLOSSA

1a With elongate papillae on the dorsum; without parapodia capable of covering the dorsum 2
1b Without elongate papillae on the dorsum; with well-developed parapodia that may cover the dorsum or be held flat against the substrate *Elysia hedgpethi* Marcus, 1961
2a Elongate papillae (about 10) restricted to the posterior half of the dorsum; most of body black or dark grey (head and tail region pale grey) *Olea hansineensis* Agersborg, 1923
2b Elongate papillae (usually more than 10) on both anterior and posterior halves of the dorsum; body usually not black or dark grey (except in *Hermaeina smithi*) 3
3a Rhinophores absent; head with short dorsolateral tentacles; edges of foot extended laterally well beyond the body; body greenish, with black streaks *Alderia modesta* (Lovén, 1844)
3b Rhinophores present; head either without dorsolateral tentacles or with very short tentacles (not to be confused with the rhinophores, which arise in the same general area); foot not extended laterally much beyond the body 4
4a Rhinophores solid in cross-section; body translucent white with an intricate reddish brown pattern on the dorsum and cerata *Stiliger fuscovittata* Lance, 1962
4b Rhinophores rolled into a cylinder throughout their length, with the edges slightly overlapping; body not covered with a reddish brown pattern 5
5a Basal portion of rhinophores rolled up, the distal portion simple; body black, with white areas around the mouth, anterior edge of the foot, tips of the cerata, and inside of the rhinophores; cerata long and bulbous *Hermaeina smithi* Marcus, 1961
5b Rhinophores rolled up throughout their length; body white, with a green or brown pattern on the dorsum and cerata; cerata long and slender 6
6a Body white, with green markings; cerata numerous (up to 50 or more) *Hermaea dendritica* (Alder and Hancock, 1843)
6b Body white, with blackish brown markings; with about 18 to 24 cerata *Hermaea ?vancouverensis* O'Donoghue, 1924

ORDER NUDIBRANCHIA

1a Gills and anus located on the midline of the dorsum, one-half to three-fourths of the body length from the anterior end; gills in a circle or semi-circle around the anus Suborder Doridacea (in part)
1b Gills and anus not located on the midline of the dorsum (either located at the posterior end of the body, on the right side, or on the dorsum near the right side); gills, if present, arranged in a row on both sides of the body 2
2a Anus at the posterior end of the body, on the midline, and located in a mantle groove between the edge of the dorsum and the foot; dorsum without dorsal processes other than rhinophores; dorsum smooth Suborder Doridacea (in part)

2b Anus on the right side of the body, or on the dorsum near the right side; usually with dorsal processes (simple or branched cerata) as well as rhinophores (if dorsal processes are not present, the dorsum has several white longitudinal ridges) 3
3a Without dorsal processes other than the rhinophores; dorsum with several white longitudinal ridges on a pinkish brown background Suborder Arminacea (in part)
3b With dorsal processes (cerata or branchial tufts) in addition to the rhinophores; without white longitudinal ridges on the dorsum 4
4a Clavus of the rhinophores partially or fully retractible into the sheath; cerata either club-shaped (with rings of large tubercles), dendritic, paddlelike, or small branchial tufts; with numerous branchial plumes or 4 to 10 pairs of cerata arranged in 2 longitudinal rows on the dorsolateral margins of the body Suborder Dendronotacea
4b Clavus of the rhinophores not retractible into the sheath (a rhinophore sheath is not present, and the rhinophore stands free); cerata slender and cylindrical, lanceolate, or leaflike; cerata usually numerous (more than 20) and usually arranged in transverse rows on the dorsum 5
5a Anus on a relatively large, conspicuous papilla near the right side of the posterior third of the dorsum, usually among the posterior third of the cerata Suborder Arminacea (in part)
5b Anus on a relatively small, inconspicuous papilla near the right side of the anterior half of the dorsum, usually among the anterior third of the cerata Suborder Eolidacea

SUBORDER ARMINACEA

1a Without cerata; dorsum with several white longitudinal ridges on a pinkish brown background (these ridges converge anteriorly toward the rhinophores) *Armina californica* (Cooper, 1862)
1b Cerata present; dorsum without white longitudinal ridges 2
2a Cerata slender, circular or only slightly oval in cross-section, and approximately the same diameter over most of their length; without a prominent oral veil 3
2b Cerata leaflike (broad in the middle and tapering toward either end), distinctly flattened in cross-section; with a prominent oral veil on the head 4
3a Body translucent white; with an irregular dorsomedial red band (this may be interrupted) running from in front of the rhinophores to near the last pair of cerata; crest between the rhinophores distinctly red; tips of cerata with a proximal orange ring followed by a transparent colorless ring and finally by an opaque white tip *Antiopella fusca* (O'Donoghue, 1924)
3b Body translucent, white, with tiny (seen with dissecting microscope) blue spots scattered on the dorsum; without a dorsomedial red line; crest between the rhinophores pale orange at most; tips of cerata with a proximal orange band succeeded by white, or with an orange band succeeded by white and finally blue bands *Antiopella barbarensis* (Cooper, 1863)
4a Body orange, with white splotches on the dorsum; with white lines on the edges of the cerata, but not around the foot or oral veil, or between the rhinophores *Dirona aurantia* Hurst, 1966
4b Body translucent, white, rarely mauve (occasionally the cerata are

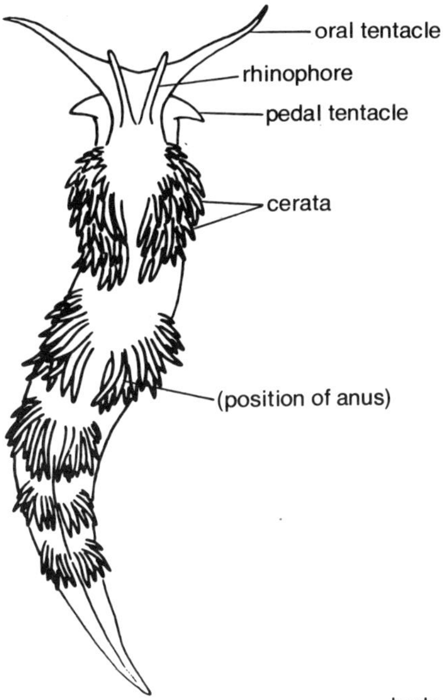

Hermissenda
ORDER EOLIDACEA

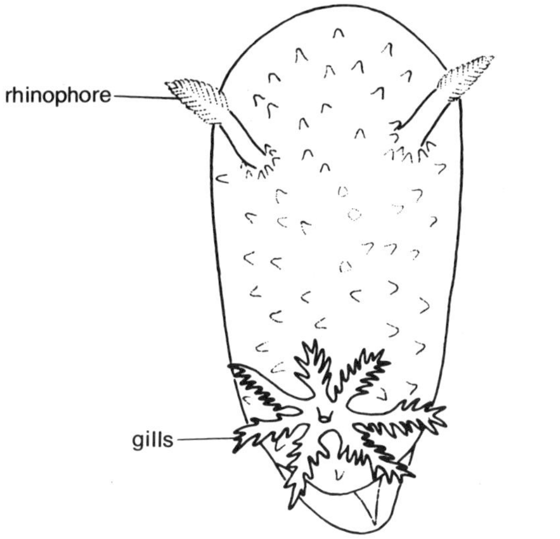

Acanthodoris
ORDER DORIDACEA

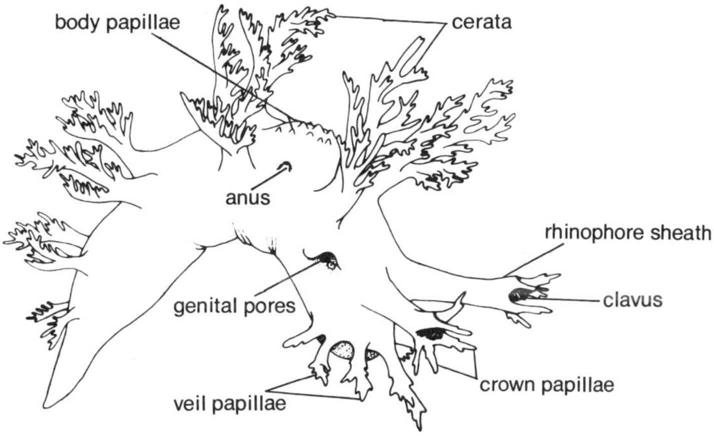

Dendronotus
ORDER DENDRONOTACEA

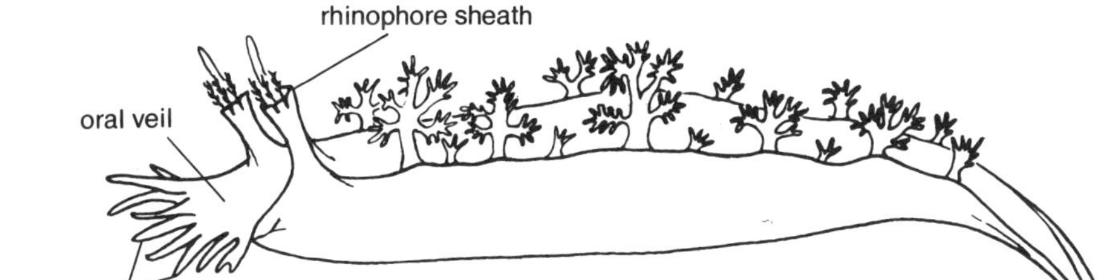

Tritonia
ORDER DENDRONOTACEA

pinkish); with white lines on the edges of the cerata, around the edge of the foot and oral veil, and between the rhinophores
Dirona albolineata Cockerell and Eliot, 1905

SUBORDER DENDRONOTACEA

1a Body stout (much more massive than most slugs); dorsal processes represented by numerous small branchial tufts concentrated along a distinct dorsolateral margin on each side of the body 2

1b Body sluglike; dorsal processes represented by cerata that are either club-shaped (with rings of large tubercles), dendritic, or paddlelike, these cerata being arranged in a longitudinal row on each side of the dorsum (but there are no distinct dorsolateral margins) 5

2a With white branchial tufts in unbroken rows extending from the rhinophores to the tail; anterior border of the oral veil white, crenulate, but lacking papillae; body apricot-orange on the dorsum (with white patches of pigment), lighter on the sides
Tochuina tetraquetra (Pallas, 1788)

2b With noticeable spaces between the branchial tufts of each row; anterior border of the oral veil with simple papillae; body pale pink to deep salmon or rose 3

3a Body relatively long and narrow; background color translucent white, with pale pink on the dorsum; usually with an intricate pattern of white lines on the dorsum; edge of rhinophore sheath smooth; with about 10 long, slender, simple papillae on the oral veil
Tritonia festiva (Stearns, 1873)

3b Body relatively stout; background color salmon or rose, darkest on the dorsum and lightest on the sides; without an intricate pattern of white lines on the dorsum; edge of rhinophore sheath crenulate; with 10 to 30 short, simple papillae on the oral veil 4

Although external characters aid in separating the two species of *Tritonia* below, dissection may be necessary to confirm identification.

4a Body about as wide as high; length up to 8 cm; with up to 25 papillae on the oral veil, but usually 10 to 16. (Dissection shows a white oral canal, squat penis, and denticles on the masticatory border.)
Tritonia exsulans Bergh, 1894

4b Body decidedly wider than high; length up to 12 cm; usually with 18 papillae on the oral veil. (Dissection shows a reddish brown or orange oral tube, an elongate, clubbed penis, and absence of denticles on the masticatory border.) *Tritonia diomedea* Bergh, 1894

5a With a large oral hood with numerous filiform tentacles around the margin; usually with 4 to 6 pairs of paddle-shaped cerata; with small earlike rhinophores on the oral hood (body translucent yellowish gray, with opaque brown hepatic diverticula) *Melibe leonina* (Gould, 1853)

5b Without an oral hood, although an oral veil may be present; cerata dendritic or club-shaped; rhinophores large and distinct 6

6a Cerata club-shaped, with 3 to 6 whorls of large, rounded tubercles; clavus of rhinophores smooth; rhinophore sheath may be crenulate, but not extended as crown papillae 7

6b Cerata dendritic; clavus of rhinophores perfoliate (with encircling flanges); rhinophore sheath extended into 4 to 10 simple or branched crown papillae 8

7a Body white, with dark brown or black penciling on the dorsum; with black rings around the bases of most or all of the tubercles on the cerata *Doto columbiana* O'Donoghue, 1921

7b Body white, with or without black penciling on the dorsum; without black rings around the bases of the tubercles on the cerata *Doto* spp. (May include 2 or more poorly described species, including *D. kya* and *D. amyra.)*

8a Large, wide, horseshoe-shaped oral veil overhanging the mouth region; usually with 3 pairs of long, branched veil papillae on the anterolateral third of the veil; with a white line around the edge of the foot, and white or yellowish white spots liberally scattered over the whole body, including the cerata, rhinophores, and veil papillae (background color pale red-brown, with darker reddish brown patches)
Dendronotus albopunctatus Robilliard, 1972

8b Veil not large and wide, and only slightly overhanging the mouth region; if only 3 pairs of veil papillae are present, these are seldom long and branched; usually without a white line around the edge of the foot, but if such a line is present, there are few or no small white spots scattered over the body 9

9a With a vertical row of 3 to 6 small, dendritic processes on the posterior border of the rhinophore stalk; with an opaque white line around the edge of the foot (body color gray or gray-brown to deep orange-red; with 4 to 7 pairs of tall, dendritic cerata tipped with dark purple, orange, white, or yellow, or some combination of these colors; stout, sluglike body up to 30 cm long [usually 6 to 18 cm])
Dendronotus iris Cooper, 1863

9b Without processes on the posterior border of the rhinophore stalk; without an opaque white line around the foot 10

10a Without lateral papillae on the rhinophore stalk; cerata tall and stout, branching in a rosette pattern about halfway up; body usually pale to dark reddish brown, mottled, and with 4 parallel longitudinal lines of reddish brown on the dorsum, 2 on each side, joining the bases of the rhinophores and each successive pair of cerata (occasionally, the body is white, orange, or yellow, but at least portions of these 4 reddish brown lines will be evident) (length up to 5 cm)
Dendronotus subramosus MacFarland, 1966

10b Usually with lateral papillae on the rhinophore stalks (however, lateral papillae may be reduced or absent in *D. albus* and *D. diversicolor*); cerata branched in a fan-like pattern, with the fan oriented at right angles to the body; without 4 parallel longitudinal lines of reddish brown on the dorsum 11

11a Body relatively stout, the length usually over 4 cm, and sometimes up to 30 cm; with 6 to 8 pairs of highly dendritic cerata; with 4 to 5 pairs of large, extensively branched veil papillae; with 20 or more simple or branched lip papillae 12

11b Body relatively delicate, the length usually 1 to 4 cm, but sometimes up to 6 cm; with 4 to 8 pairs of rather sparsely branched cerata; with 3 to 4 pairs of veil papillae, these occasionally extensively branched; with up to 8 simple lip papillae 13

12a With a red line around the foot; body either grayish white, with magenta pigment on the distal portion of the cerata, rhinophores, veil, lip, and lateral papillae, or, more rarely, mottled brownish-red and pale pink all over; with 6 to 8 pairs of tall, extensively branched cerata (the ceratal branches long); with numerous small accessory cerata between the principal cerata; anterior lip papillae large and branched, but the posterior and ventral lip papillae usually simple;

with 5 pairs of extensively branched veil papillae; with 5 pairs of extensively branched crown papillae on the rhinophores; body up to 28 cm long, usually 8 to 15 cm long *Dendronotus rufus* O'Donoghue, 1921

12b Without a red line around the foot; body translucent white to pink, with opaque white pigment on the distal third of the cerata and rhinophores; with 6 to 7 pairs of short, extensively branched cerata (the branches relatively short), without accessory cerata between the principal cerata; with numerous (often 30 or more) small, simple, fingerlike lip papillae; with 4 or 5 pairs of short, moderately branched veil papillae; 4 to 12 (usually 5) moderately branched crown papillae on the rhinophores, but not necessarily the same number on each rhinophore; body up to 14 cm long, usually 4 to 8 cm long *Dendronotus dalli* Bergh, 1879

13a Body white or lilac, with white and/or orange on the tips of the cerata; with a dorsomedial white line extending from the tip of the tail to at least the posteriormost pair of cerata 14

13b Body pale brown to dark reddish brown, sometimes whitish; without a dorsomedial white line between the tip of the tail and the posteriormost pair of cerata. (The following color morphs prevail: 1) body pale brown, with yellow caps on the tubercles on the body, and with small yellow or white flecks on the processes of the dorsum; 2) body whitish, with a few small pinkish brown spots on the dorsum; 3) body dark reddish brown, often with large patches of white pigment between the cerata and/or on the sides, ventral to the cerata, and with white or yellow spots on the dorsum and processes. All color morphs have 5 to 7 pairs of moderately branched cerata; 4 pairs of simply branched veil papillae; up to 6 simple lip papillae; a relatively delicate sluglike body; length up to 5 cm, usually 1.5 to 3 cm) *Dendronotus frondosus* (Ascanius, 1774)

14a With 5 to 8 pairs of moderately branched cerata; often with a few subconical papillae on the dorsum; body translucent white; with opaque white pigment on the tips of the cerata, crown, veil papillae, and lateral papillae; with a dorsomedial white line extending from the tip of the tail to the level of the 4th to 2nd pairs of cerata; in some individuals, the core of cerata grades (proximally to distally) from tan to golden yellow to copper to white; length up to 4.5 cm, usually 1 to 3 cm *Dendronotus albus* MacFarland, 1966

14b Usually with 4 pairs (rarely 5 pairs) of tall, slender, sparsely-branched cerata; body without papillae; body translucent white or lilac, with intergrades occasionally seen; with opaque white and/or metallic orange pigment on the tips of the cerata, rhinophores, and veil papillae; with a dorsomedial white line extending from the tip of the tail to the posteriormost pair of cerata; length up to 7 cm long, usually 3 to 6 cm *Dendronotus diversicolor* Robilliard, 1970

SUBORDER EOLIDACEA

The species dealt with in couplets 9 to 15 are difficult to key because they are very similar in external morphology. Although color is used to differentiate these species, it varies to some extent. For certain identification, it is advisable to consult the original descriptions or other extensive descriptions, as those provided by Roller (1969, 1972) and Edmunds & Kress (1969).

1a At least 3 or 4 rows of cerata on each side originating along the anterior and lateral margins of the head, anterior to the bases of the rhinophores 2
1b All rows of cerata originating posterior to the bases of the rhinophores 4
2a Longest cerata more than half the length of the body; cerata usually lie flat and are oriented approximately at right angles to the anterior-posterior axis of the body; foot proportionately broad (usually less than 3 times as long as wide), tapering to a point at the tail *Cumanotus beaumonti* (Eliot, 1906)
2b Longest cerata less than one-third the length of the body; cerata usually lie nearly flat on the dorsum and are oriented approximately parallel to the anterior-posterior axis of the body; foot relatively narrow (usually more than 4 times as long as wide), bluntly rounded at the tail 3
3a Anterolateral corners of foot projecting as acute pedal tentacles; body translucent white and usually covered with gray to brown spots, these being most dense on the cerata (some specimens, especially from floats, lack any brown or gray pigment); usually with a prominent triangular white patch anterior to the rhinophores; body not translucent enough to show hepatic diverticula or gonads; tips of cerata may be opaque white; usually associated with anemones, especially *Metridium* and *Anthopleura* *Aeolidia papillosa* (Linnaeus, 1761)
3b Anterolateral corners of foot rounded and not projecting laterally; body translucent cream with a tinge of pink; no white patch anterior to the rhinophores; hepatic diverticula purple-gray to rose in color; gonads rose pink; tips of cerata opaque white; usually associated with hydroids *Precuthona divae* Marcus, 1961
4a Anterolateral corners of foot elongated as acute, relatively prominent pedal tentacles 5
4b Anterolateral corners of foot bluntly rounded 9
5a Background color purple; rhinophores and cerata bright orange over their whole length *Flabellinopsis iodinea* (Cooper, 1862)
5b Background color not purple, usually translucent white; rhinophores and cerata not bright orange over their whole length (however, they may be partly orange) 6
6a With an orange band beginning just anterior to the rhinophores and passing between them, reaching the first group of cerata (a similar orange band may appear over the cardiac region); orange areas usually bounded by broad, opaque white or "electric" blue dorsomedial lines which begin on the oral tentacles and continue to the tip of the tail; tips of cerata usually orange or chrome yellow, with some white at the extremity *Hermissenda crassicornis* (Eschscholtz, 1831)
6b Without orange bands on the dorsum; white line(s) on the dorsum narrow and may be broken; tips of cerata opaque white, but rarely orange 7
7a With 3 narrow, longitudinal white lines on the body: 1 dorsomedial, running from the tip of the tail to a level just anterior to rhinophores, where it splits, 1 branch going to the tip of the oral tentacle on each side; 1 on either side just ventral to the base of the cerata, beginning at the first group of cerata and joining the dorsomedial line posterior to the last group of cerata *Coryphella trilineata* O'Donoghue, 1921
7b With a single dorsomedial white line, broken here and there and often restricted to the tail region, but without lateral white lines 8

8a Oral veil between the oral tentacles usually projecting dorsally as a blunt, triangular, median papilla; short, blunt tail extending just past the last group of cerata; pedal tentacles long; hepatic diverticula thin (less than one-third the width of the cerata); white dorsomedial line extending from the tip of the tail to the most posterior group of cerata only; distal one-fourth of each ceras opaque white *Coryphella fusca* O'Donoghue, 1921

8b Oral veil slightly rounded between the oral tentacles, not projecting dorsally; long, slender, acute tail extending one-third to one-half of the total body length past the last group of cerata; pedal tentacles short; hepatic diverticula at least three-fourths the width of the cerata; white dorsomedial line extending from the tip of the tail to the rhinophores, although often broken; only the tips of the cerata opaque white *Coryphella longicaudata* O'Donoghue, 1922

9a Surface of cerata slightly bumpy; cerata usually held erect; cerata cylindrical, and slightly inflated in the central portion; cerata often shed when the animal is disturbed; all cerata joined together by an irregular network of hepatic diverticula 13

9b Surface of cerata usually smooth; cerata may be held down on the body; cerata usually not inflated or only slightly so, although in some species they may be rather flat, tapering from a wide central portion to a bluntly rounded tip; cerata not readily shed; cerata joined by the hepatic diverticula in a neat symmetrical pattern, usually in transverse rows running across the body 10

10a With a purple band encircling each rhinophore and oral tentacle near its middle (sometimes there is a white band above and/or below the purple band, and there may also be some purple at the bases of the rhinophores and oral tentacles); the proximal third of hepatic diverticula in each ceras dark red, the central third olive-green to brown, the distal third pale yellow, and the tip colorless (3 opaque white bands [solid or broken] on the cerata accentuate the color changes of the hepatic diverticula) *Trinchesia abronia* (MacFarland, 1966)

10b Without a purple band encircling each rhinophore and oral tentacle; hepatic diverticula usually of one color throughout 11

11a With a more or less solid accumulation of opaque white pigment on the rhinophores, on the dorsal side of the head anterior to the rhinophores, on the dorsum from the rhinophores to the last pair of cerata and down the sides to the lateral cerata, on the anterolateral face of the proximal portion of each of the cerata, and on the entire surface of the distal portion of the cerata (except for the tip, which is clear); hepatic diverticula may be pale reddish brown or green; no other pigmentation prominent *Trinchesia albocrusta* (MacFarland, 1966)

11b White pigment, if present, not so extensive as above, usually limited to patches on the cerata or rhinophores, and/or the dorsum; other pigmentation usually present 12

12a Tiny bright orange spots densely scattered on the proximal portions of the oral tentacles and on the distal portions of the rhinophores; opaque white usually on the tips of the rhinophores, on the dorsal side of the oral tentacles (overlain by orange spots), on the anterior face of the cerata, and sometimes on the dorsal side of the head anterior to the rhinophores; hepatic diverticula usually bright red (although they may be yellowish; the color depends on the prey) *Catriona alpha* (Baba & Hamatani, 1963)

12b No tiny bright orange spots densely scattered on both the proximal

portions of the oral tentacles and the distal portions of the rhinophores (however, *Trinchesia lagunae* (O'Donoghue, 1926), if it should be found in our area, has orange spots on the rhinophores)
Various unidentified eolids

Included here are a number of small and incompletely understood eolids of various genera (*Trinchesia, Cratena, Catriona, Cuthona*). Studies of systematic and ecological relationships are needed to determine what species are present. It is likely that there are some undescribed or poorly described species involved.

13a Body pale translucent yellowish green, with fine specks of brown, black, and white; hepatic diverticula in the cerata and those running between the bases of the cerata greenish in color; usually with fewer than 15 cerata; associated with hydroids of the genus *Obelia* (especially *O. longissima* and *O. commissuralis*)
Eubranchus olivaceus (O'Donoghue, 1922)

13b Body translucent white; sometimes with rust-brown splotches on the dorsum and cerata, and small white spots on the cerata, rhinophores, and oral tentacles; hepatic diverticula in the cerata and those running between the bases of the cerata not greenish in color (usually reddish or yellowish brown); usually with more than 15 cerata; not associated with *Obelia* 14

14a Body translucent white; opaque white on the tips of the cerata and often on the dorsum anterior to the heart region; hepatic diverticula red or reddish tan, giving the animal as a whole a reddish hue; usually associated with the hydroid *Sertularella tricuspidata*
Eubranchus sanjuanensis Roller, 1972

14b Body translucent white, but covered to varying degrees with splotches of rust-brown pigment; opaque white on the tips of the rhinophores and cerata, and in a ring near the middle of each of the cerata (distal to the white ring is a subterminal ring of rust-colored spots; there is also a ring of rust-colored spots near the middle or beginning of the distal quarter of each oral tentacle and rhinophore); hepatic diverticula brown, sometimes with yellow spots; associated with the hydroid *Plumularia lagenifera* *Eubranchus rustyus* (Marcus, 1961)

SUBORDER DORIDACEA

1a Anus on the midline at the posterior end of the body, situated in a mantle groove between the foot and edge of the dorsum; gills located on either side of the anus; dorsum lacking processes other than the rhinophores; pale ochre spots forming an intricate pattern on the otherwise white dorsum, and animals closely resemble the bryozoan *Membranipora*, with which they are usually associated 2

1b Anus on the midline of the dorsum, one-half to one-fourth of the body length from the anterior end; gills in a circle or semicircle around the anus; dorsal processes include gills and rhinophores, and may also include various other papillae or tubercles; body may be white, but an intricate pattern of ochre spots is absent and the animals are neither regularly associated with *Membranipora* nor do they resemble it 3

2a With a distinct notch on the psoterior border of the dorsum; with 6 to 14 gills on either side of the medial anal opening
Corambe pacifica MacFarland and O'Donoghue, 1929

2b Margin of the dorsum unbroken, not notched at the posterior end; with 2 to 4 gills on either side of the medial anal opening

Doridella steinbergae (Lance, 1962)

3a Sluglike; dorsum usually with some conspicuous large processes in addition to rhinophores and gills, these processes usually near the rhinophores or gills or on the dorsolateral border of the body 4

3b Doridiform; dorsum without large processes other than rhinophores and gills, although there may be numerous low tubercles or nearly microscopic slender papillae covering the whole dorsum 10

4a Body firm due to spicules in the dorsum; with short, blunt, cylindrical papillae arranged in 4 to 6 irregular longitudinal rows on the dorsum; body white to cream, with brown to black spots scattered on the dorsum *Aegires albopunctatus* MacFarland, 1905

4b Body relatively soft, without a firm network of spicules in the dorsum; without short cylindrical papillae arranged in irregular longitudinal rows (although there may be some short or long papillae scattered over the dorsum); body without brown to black spots on the dorsum 5

5a Body somewhat flattened dorsoventrally; with numerous (over 40) slender, club-shaped papillae all around the margin of the dorsum, and also with a few short papillae in the central area of the dorsum; body white or cream, with bright orange on the tips of the papillae and on the rhinophores, but not on the gills and not in spots on the dorsum itself *Laila cockerelli* MacFarland, 1905

5b Body not flattened dorsoventrally; with fewer than 30 papillae of various shapes around the margins of the dorsum, but none of these papillae are slender and club-shaped; bright orange, if present in the color pattern, found on the gills and in spots on the dorsum itself, as well as on the papillae and rhinophores 6

6a With a few (about 10 to 15) small coronate papillae along each dorsolateral margin of the dorsum; with about 8 to 15 small irregularly papillated processes on the edge of the oral veil; body white to pale yellow, with deep orange on the tips of the gills, on the rhinophores, tail, dorsal papillae, and oral veil papillae, and also forming spots on the dorsum itself *Triopha carpenteri* (Stearns, 1873)

6b Without coronate papillae on the dorsolateral border or on the oral veil; with little or no orange color anywhere 7

7a With long, cylindrical, tapered papillae on the dorsolateral margin between the rhinophores and gills, and on central area of the dorsum anterior to the gills; with 14 pinnate gills; color of living animal not known

Cargoa vancouverensis (O'Donoghue, 1921) *(=Okenia vancouverensis)*

7b If long, cylindrical, tapered papillae are present, they occur only at the edge of the oral veil, lateral to the gills, and immediately anterior to the rhinophores (papillae are not usually found on the dorsolateral margin between the rhinophores and gills, or on the central area of the dorsum anterior to the gills; however, *Polycera zosterae* may have short, blunt papillae scattered over much of the body); with fewer than 8 pinnate gills 8

8a Background color white or cream, with markings of yellow or of both yellow and black 9

8b Body deep olive to black *Polycera zosterae* O'Donoghue, 1924

9a With only yellow pigment on the tips of all papillae, gills, and rhinophores; with 3 parallel yellow lines between the rhinophores and gills; with 2 long, simple papillae on either side of the body just anterior to the rhinophores; with 3 or 4 erect papillae on either side of the gills; with 2 papillae on the right and left edges of the oral veil *Ancula pacifica* MacFarland, 1905

9b On all papillae, gills, and rhinophores, proximally to distally, the color changes from black to yellow to white; with yellow spots scattered on the dorsum, and with a yellow line on the edge of the foot; without papillae just anterior to the rhinophores; with 3 to 6 erect papillae on both sides of the gills; with 6 to 12 papillae on the right and left edges of the oral veil *Polycera tricolor* Robilliard, 1971

10a Gills completely retractable into a sheath; rhinophores short and blunt 11

10b Gills not completely retractable into a sheath; rhinophores long, tapering 20

11a Without large, rounded tubercles on the dorsum; the entire surface either nearly smooth (tiny tubercles visible under dissecting microscope) or minutely villous 12

11b With both large and small, rounded tubercles scattered over the dorsum, but the surface otherwise smooth 13

12a Body white to dusky gray; spots up to 1.5 cm in diameter, scattered on the dorsum, these spots usually composed of a concentric white ring enclosing brown inner rings with dark brown to black centers *Diaulula sandiegensis* (Cooper, 1862)

12b Body off-white to pale yellow, but seldom dusky gray; without spots composed of concentric white and brown circles (however, small black spots, less than 1 mm in diameter, are scattered over the dorsum) *Discodoris heathi* MacFarland, 1905

13a Body red or deep orange-red; a few tiny black spots may be present on the dorsum; rhinophores with 10 to 12 vertically oriented leaves, and terminating in simple slender papillae *Rostanga pulchra* MacFarland, 1905

13b Body white, yellow, or pale orange; yellow or black spots may be present on the dorsum, but they may be lacking; rhinophores perfoliate (the leaves oriented horizontally, as typical for dorids) 14

14a Body either yellow (varying from lemon yellow to dusky greenish yellow) or pale orange; with black spots on the dorsum; without any distinct pigment on the edge of the dorsum 15

14b Body white to cream, sometimes with yellow spots on the tubercles on the dorsum; without black spots on the dorsum; dorsum may be edged with yellow 17

15a With black spots (up to 11), usually arranged in a single row on the dorsal midline, from just anterior to the rhinophores to the gills; body chrome yellow or pale orange *Aldisa sanguinea cooperi* Robilliard and Baba, 1972

15b With black spots (usually more than 11) scattered over the entire dorsum; body color lemon yellow to dusky greenish yellow, rarely with an orange tinge 16

16a Body yellow to orange-yellow; black spots usually present between the tubercles only, rarely on them (in unusual specimens, black spots are absent altogether); gills and rhinophores usually paler than the body; exudes considerable quantities of "sickly sweet-smelling" mucus when disturbed *Anisodoris nobilis* (MacFarland, 1905)

16b Body lemon yellow to greenish yellow, rarely orange, with black spots regularly on the tubercles as well as between them; gills and rhinophores usually duskier than the body; exudes little mucus and has almost no odor *Archidoris montereyensis* (Cooper, 1862)

17a Entire body white (or rarely, pale custard), without any clear yellow spots or lines; dorsum high, arched *Archidoris odhneri* (MacFarland, 1966)

17b Body mostly white, but with yellow spots on the dorsum and/or a yellow line around the margin of the dorsum (in addition, the clavus of the rhinophores usually has some yellow to dark brown pigment); dorsum rather flat 18

18a With a solid lemon-yellow line around the margin of dorsum and on the posterior margin of foot; with yellow tips on the gills and on most of the dorsal tubercles; rhinophores usually white, except for yellow at their tips *Cadlina luteomarginata* MacFarland, 1966

18b Without a solid yellow line around the margin of the dorsum or foot; without yellow on the tips of the gills, and the yellow spots on the dorsum are generally large and mostly near the margin; rhinophores dusky brown to almost black 19

19a Rhinophore leaves brown to almost black; with minute brown spots scattered along the labial tentacles, and with microscopic brown spots on the dorsum *Cadlina flavomaculata* MacFarland, 1905

19b Rhinophore leaves pale brown; without minute brown spots on the labial tentacles, and without microscopic brown spots on the dorsum or elsewhere *Cadlina modesta* MacFarland, 1966

20a Dorsum covered with long, slender, flexible, conical or sub-conical, close-set papillae (surface "fuzzy"); without spicules protruding through the tips of the papillae 21

20b Dorsum with stout, rounded, conical to sub-conical or mushroom-shaped tubercles (surface "firm"); spicules often projecting through the tips of the tubercles 23

21a Body all white or just slightly off-white; no other color on the dorsum or dorsal processes *Acanthodoris* sp. (undescribed)

21b Background color white or some shade of gray; yellow, gray, or port-wine red (or some combination of these colors) present on the dorsum or dorsal processes 22

22a Background color white; with a solid lemon-yellow line around the margin of the dorsum; without gray on the dorsum; with yellow on the tips of the rhinophores and gills, and on the dorsal papillae; usually with 5 gills *Acanthodoris hudsoni* MacFarland, 1905

22b Background color white to mauve-gray (all intergrades seen); without a yellow line around the margin of the dorsum; with dark gray splotches on the dorsum of darker animals; with port-wine red on the tips of the rhinophores and gills (but this is difficult to see on darker specimens), and with yellow on the tips of the dorsal papillae; usually with 9 gills *Acanthodoris nanaimoensis* O'Donoghue, 1921

23a Body dull cream with irregular brown bands on the dorsum, or brown in general, usually at least 1.5 cm long 24

23b Body usually white, cream, or pale orange, without any brown; usually less than 1.5 cm long 25

24a Gills unipinnate, almost erect, 16 or more in number, arranged in a broad horseshoe around the anus, with the opening of the horseshoe directed posteriorly; papillae mushroom-shaped, with spicules protruding from their tips; body usually dull cream, with darker brown pigment in 2 or 3 irregular longitudinal bands on the dorsum; without yellow on the gills, papillae, rhinophores, or around the edge of the foot *Onchidoris bilamellata* (Linnaeus, 1767)

24b Gills bipinnate, spreading outward when expanded, usually 5 to 7 in number, completely encircling the anus; most papillae conical or subconical, almost none of them with protruding spicules; body pale brown, with small splotches of black scattered between the tubercles; with yellow on the tips of the gills, body papillae, rhinophores, and often around the edge of foot *Acanthodoris brunnea* MacFarland, 1905

25a Many or all tubercles on the dorsum mushroom-shaped; bundles of spicules sometimes protruding through the distal ends of the tubercles; spicules in the dorsum tending not to radiate from the bases of the tubercles, but rather to form parallel bundles running between the tubercles; usually with a distinct, large, white tubercle immediately posterior to the anus 26

25b Few or none of the tubercles on the dorsum mushroom-shaped (most are sub-conical, inflated, and pointed at the tip); bundles of spicules protruding through the distal ends of the tubercles; large spicules in the dorsum tending to radiate from the bases of the tubercles; without a distinct, large, white tubercle posterior to the anus *Adalaria* sp. (undescribed)

26a Usually more than 1.5 cm long; usually found in close association with barnacles, upon which it feeds (brown-mottled specimens of the species [choice 24a, above] often occur in the same population) *Onchidoris bilamellata* (Linnaeus, 1767)

26b Usually less than 1.5 cm long; usually found in close association with *Schizoporella* and similar bryozoans, which are its prey *Onchidoris* sp. (undescribed species)

The following species of nudibranchs may occur in our area, but they have not come to the attention of the author. Moreover, the original descriptions of some of them are so inadequate that they would be unsuitable for confirming identifications. Some of the names may be junior synonyms. Further work will be required to resolve the taxonomic difficulties involved.

ORDER DENDRONOTACEA
- *Doto wara* Marcus, 1961
- *Doto kya* Marcus, 1961
- *Doto ganda* Marcus, 1961
- *Doto amyra* Marcus, 1961

ORDER EOLIDACEA
- *Cuthona concinna* (Alder and Hancock, 1843)
- *Cratena columbiana* (O'Donoghue, 1922)
- *Coryphella rufibranchialis* (Johnston, 1832)

ORDER DORIDACEA
- *Acanthodoris atrogriseata* O'Donoghue, 1927
- *Acanthodoris pilosa* (Abildgaard, 1789)
- *Acanthodoris armata* O'Donoghue, 1927
- *Onchidoris muricata* (Müller, 1776)
- *Onchidoris hystricina* (Bergh, 1878)
- *Onchidoris fusca* (Müller, 1776)
- *Cargoa plana* (Baba, 1960) *(=Okenia plana)*
- *Doris o'donoghuei* Steinberg, 1963

Baba, K., & I. Hamatani. 1963. A cuthonid, *Cuthona alpha* n. sp., with a radula of *Catriona*-type (Nudibranchia-Eolidacea). Publ. Seto Mar. Biol. Lab., 11:330-43.

Beeman, R. O. 1968. The order Anaspidea. Veliger, 3 (suppl., pt. 2):87-102.

Bergh, L. S. R. 1879. On the nudibranchiate gastropod Mollusca of the north Pacific Ocean, with special reference to those of Alaska. Part I. Proc. Acad. Nat. Sci. Philadelphia, 1879:71-132.

------. 1880. On the nudibranchiate gastropod Mollusca of the north Pacific Ocean, with special reference to those of Alaska. Part II. Proc. Acad. Nat. Sci. Philadelphia, 1880:40-127.

------. 1894. Die Opisthobranchien. *From* Reports on the dredging operations off the west coast of Mexico and in the gulf of California in charge of Alexander Agassiz, carried on by the U. S. Fish Commission Steamer "Albatross" during 1891. Bull. Mus. Comp. Zool. Harvard Coll., 25:125-233.

Burn, R. 1968. *Archidoris odhneri* (MacFarland, 1966) comb. nov., with some comments on the species of the genus on the Pacific coast of North America. Veliger, 11:90-2.

Cockerell, T. D. A. 1915. The nudibranch genus *Triopha* in California. Jour. Entom. Zool. Pomona College, 8:228-9.

Dall, W. H. 1871. Descriptions of sixty new forms of molluscs from the west coast of North America and the north Pacific Ocean, with notes on others already described. Amer. Jour. Conchol., 7:93-160.

Dall, W. H., & P. Bartsch. 1909. A monograph of west American pyramidellid mollusks. Bull. 68, U. S. Nat. Mus. xii + 258 pp.

Edmunds, M., and A. Kress. 1969. On the European species of *Eubranchus* (Mollusca: Opisthobranchia). Jour. Mar. Biol. Assoc. United Kingdom, 49:879-912.

Hurst, A. 1966. Description of a new species of *Dirona* from the north-east Pacific. Veliger, 9:9-15.

Lance, J. R. 1961. A distributional list of southern California opisthobranchs. Veliger, 4:64-9.

------. 1962. A new *Stiliger* and a new *Corambella* (Mollusca: Opisthobranchia) from the northwest Pacific. Veliger, 5:33-8.

MacFarland, F. M. 1906. Opisthobranchiate Mollusca from Monterey Bay, California and vicinity. Bull. U. S. Bur. Fish., 25:109-57.

------. 1912. The nudibranch family Dironidae. Zool. Jahrb., suppl., 15: 515-36.

------. 1925. The Acanthodorididae of the California coast. Nautilus, 39: 49-65.

------. 1926. The Acanthodorididae of the California coast. Nautilus, 39: 94-105.

------. 1966. Studies of opisthobranchiate mollusks of the Pacific coast of North America. Mem. Calif. Acad. Sci., 6. xvi + 546 pp.

MacFarland, F. M., & C. H. O'Donoghue. 1929. A new species of *Corambe* from the Pacific coast of North America. Proc. Calif. Acad. Sci., ser. 4, 18:1-27.

Marcus, E. 1961. Opisthobranch mollusks from California. Veliger, 3 (suppl.):1-84.

O'Donoghue, C. N. 1921. Nudibranchiate Mollusca from the Vancouver Island region. Trans. Royal Canad. Inst., 13:147-209.

------. 1922. Notes on the nudibranchiate Mollusca from the Vancouver Island region. III. Records of species and distribution. Trans. Royal Canad. Inst., 14:145-67.

------. 1924. Notes on the nudibranchiate Mollusca from the Vancouver Island region. IV. Additional species and records. Trans. Royal Canad.

Inst., 15:1-33.
------. 1926. A list of the nudibranchiate Mollusca recorded from the Pacific coast of North America with notes on their distribution. Trans. Royal Canad. Inst., 15:199-247.
------. 1927. Notes on the nudibranchiate Mollusca from the Vancouver Island region. V. Two new species and one new record. Trans. Royal Canad. Inst., 16:1-12.
Robilliard, G. A. 1970. The systematics and some aspects of the ecology of the genus *Dendronotus* (Gastropoda: Nudibranchia). Veliger, 12:433-79.
------. 1971. Range extensions of some northeast Pacific nudibranchs (Mollusca: Gastropoda: Opisthobranchia) to Washington and British Columbia, with notes on their biology. Veliger, 14:162-5.
------. 1971. A new species of *Polycera* (Opisthobranchia: Mollusca) from the northeastern Pacific, with notes on other species. Syesis, 4:235-43.
Roller, R. A. 1969. Nomenclatural changes for the new species assigned to *Cratena* by MacFarland, 1966. Veliger, 11:421-3.
------. 1970. A list of recommended nomenclatural changes for MacFarland's "Studies of opisthobranchiate mollusks of the Pacific coast of North America." Veliger, 12:371-4.
Sphon, G. G. 1972. Some opisthobranchs (Mollusca: Gastropoda) from Oregon. Veliger, 15:153-7.
Steinberg, J. E. 1961. Notes on the opisthobranchs of the west coast of North America. I. Nomenclatural changes in the order Nudibranchia (Southern California). Veliger, 4:57-63.
------. 1963. Notes on the opisthobranchs of the west coast of North America. II. The order Cephalaspidea from San Diego to Vancouver Island. Veliger, 5:114-7.
------. 1963. Notes on the opisthobranchs of the west coast of North America III. Further nomenclatural changes in the order Nudibranchia. Veliger, 6:63-7.
------. 1963. Notes on the opisthobranchs of the west coast of North America. IV. A distributional list of opisthobranchs from Point Conception to Vancouver Island. Veliger, 6:68-73.
Thompson, T. E. 1971. Tritoniidae from the north American Pacific coast. Veliger, 13:333-8.

SUBCLASS PULMONATA

ORDER BASOMMATOPHORA

Family Ellobiidae: *Phytia myosotis* (Draparnaud, 1801) (see key to shelled gastropods, choice 67a) is abundant in salt marshes, especially under debris and in growths of *Salicornia* at levels that are inundated only by the highest tides. Its shell is up to about 8 mm high and the periostracum is usually chestnut brown. *P. myosotis* is usually accompanied by *Assiminea californica*, a still smaller prosobranch.

Family Siphonariidae: *Siphonaria thersites* Carpenter, 1864 is a limpetlike species (see key to shelled gastropods, choice 21a) found at upper levels of the intertidal region on rocky shores.

ORDER STYLOMMATOPHORA

Family Onchidiidae. The systematic position of this group, represented locally by *Onchidella borealis* Dall, 1871, has been unstable for a long time. In this work it is referred to the subclass Opisthobranchia, order Onchidiacea (see key to orders of Opisthobranchia, choice 9a, and order Onchidiacea).

CLASS BIVALVIA

1a Body wormlike, the shell much reduced and capable of covering only a small part of it; burrowing in wood, forming long tunnels lined by a whitish calcareous deposit; siphon provided with a pair of paddlelike or nearly featherlike calcareous structures (pallets) for closing the burrow ("shipworms") 2

1b Body not wormlike, the shell (even if it is small) capable of covering at least most of it; not burrowing in wood (exception: *Xylophaga)*, although some species may nestle in holes produced by other animals; siphon not provided with calcareous pallets for closing the burrow 3

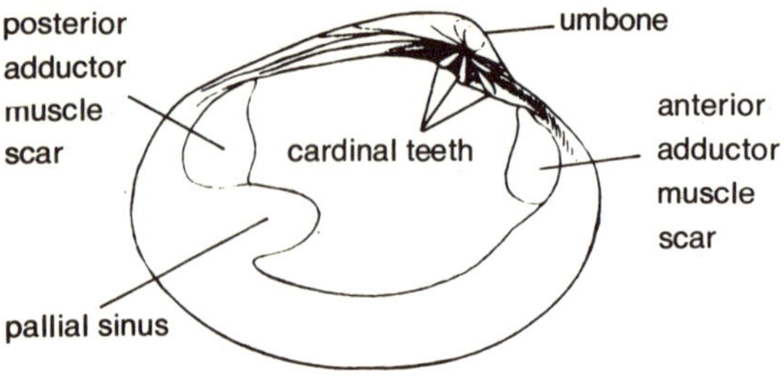

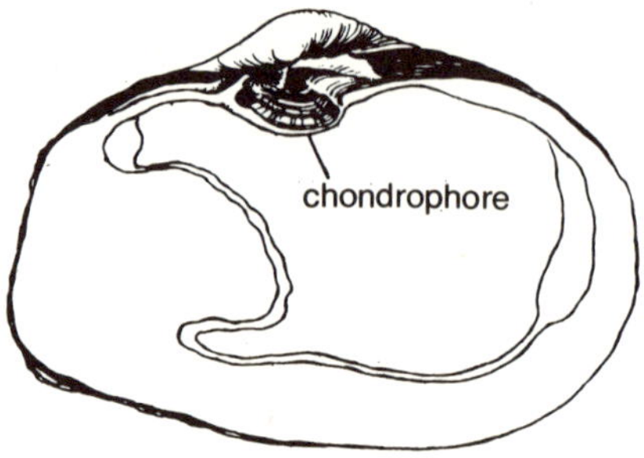

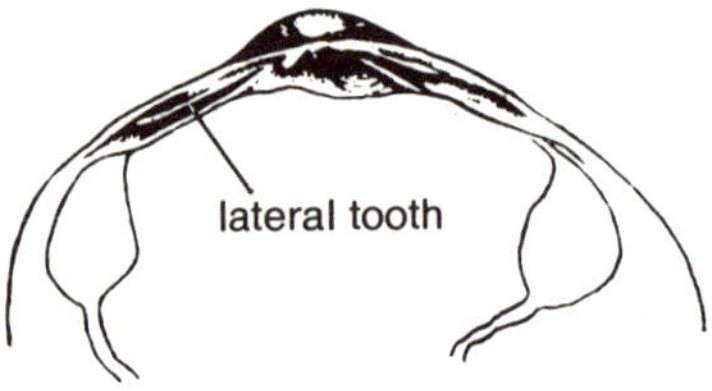

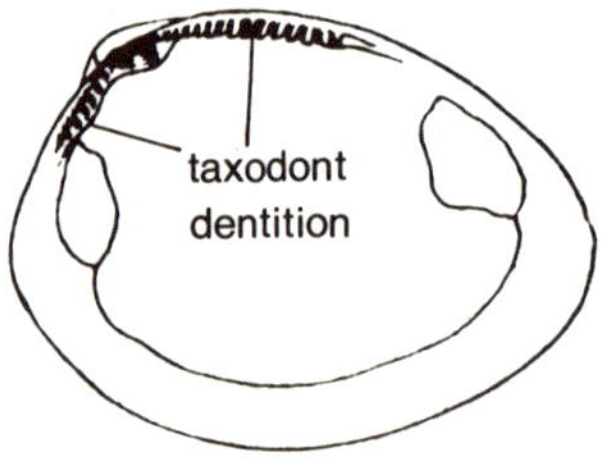

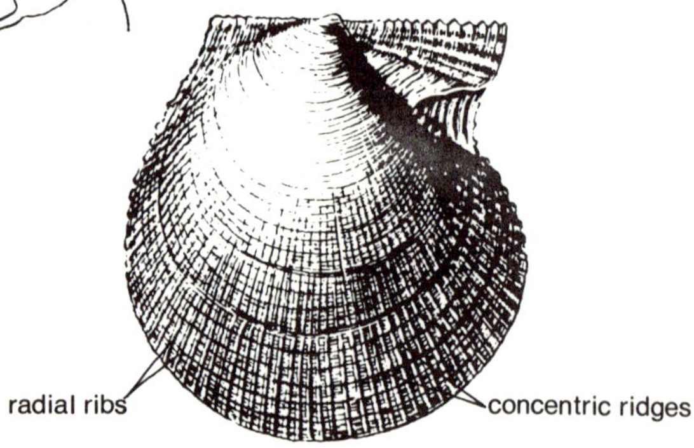

2a Pallets approximately paddlelike; rare in our area
Teredo navalis Linnaeus, 1758

2b Pallets consisting of a series of cones set one inside the other, but on the whole resembling feathers; common in our area
Bankia setacea Tyron, 1863

3a Valves of shell decidedly unequal in form as well as size; 1 valve permanently cemented to a hard substrate (rock, wood, concrete) 4

3b Valves of shell essentially similar to one another, although 1 valve may be slightly smaller or more nearly flat than the other; neither valve permanently cemented to the substrate (but the animal may be attached by a byssus of organic threads) 8

4a Attached valve with a hole or notch near the hinge *Pododesmus cepio* (Gray, 1850)

4b Attached valve without a hole or notch near the hinge 5

5a Valves very thick, sometimes almost circular in outline; with rather regularly spaced rough ridges radiating from the umbones; winglike extensions on both sides of the umbones (as in a scallop) generally persisting; interior of each valve with a conspicuous purple blotch near the hinge; with a row of eyes along the edge of the mantle *Hinnites giganteus* (Gray, 1825)

5b Oysters--valves moderately thick, generally higher than wide; without regularly spaced ridges radiating from the umbones (if ridges are present they follow no particular pattern); without winglike extensions near the umbones; interior of valves without a purple blotch near the hinge; without eyes along the edge of the mantle 6

6a Shell generally much roughened and conspicuously frilled; height up to about 25 cm *Crassostrea gigas* (Thunberg, 1793)

6b Shell not much roughened and rarely showing any frills (sculpturing usually limited to weak and irregular concentric lines); height not exceeding 15 cm 7

7a Height up to nearly 15 cm (but generally much less than this); scar of adductor muscle deep blue-black *Crassostrea virginica* (Gmelin, 1791)

7b Height up to about 5 cm; scar of adductor muscle not much darker than the rest of the interior of the valve *Ostrea lurida* Carpenter, 1864

8a Scallops--with winglike extensions on both sides of the umbone of each valve (the extension on one side may be small, however) 9

8b Without winglike extensions on both sides of the umbone of each valve 17

9a Valves capable of being closed tightly together; at least 1 of the winglike extensions on each valve conspicuous; height of shell less than 1 and one-half times the length; height frequently exceeding 2 cm 10

9b Valves gaping, not capable of being closed tightly together; winglike extensions small; height of shell less than 1 and one-half times the length; height not exceeding 1 cm *Limaria subauriculata* (Montagu, 1808)

10a Both valves decidedly convex; height less than 8 cm (except in the case of *Hinnites giganteus*, which becomes attached by one valve to rock before it reaches this size (see choice 5a) 11

10b Only the right valve decidedly convex (left valve nearly flat); height regularly exceeding 10 cm *Pecten caurinus* Gould, 1850

11a Right valve similar to the left valve in having radial ribs or "pseudoribs" (these look like ribs, but in reality consist of closely-spaced, fragile ruffles of the concentric lines of growth; if a pseudorib is abraded, it becomes evident that there is no true rib underlying it) 12

11b Right valve with only concentric growth lines, except perhaps near the margins, where faint ribs may be noticeable *Propeamussium davidsoni* (Dall, 1897)

12a With pseudoribs instead of true ribs *Propeamussium alaskense* (Dall, 1872)

12b With true ribs (though these may be delicate and no more prominent than the concentric growth lines) 13

13a Without ribs in the interior of the valves; exterior of valves with about 60 or more delicate, scaly ribs whose character can hardly be distinguished without magnification; height less than 1 cm *Cyclopecten vancouverensis* (Whiteaves, 1893)

13b With ribs in the interior of the valves; exterior of valves with ribs that can readily be seen with the unaided eye; height regularly exceeding 1 cm (at maturity, generally at least 5 cm) 14

14a Interior of valves with a purple blotch near the hinge young (1 to 2 cm) *Hinnites giganteus* (Gray, 1825)
This species eventually becomes attached by its right valve, and in time becomes considerably distorted; see choice 5a.

14b Interior of valves without a purple blotch near the hinge 15

15a Radial ribs essentially smooth (in any case not decidedly roughened by spinelike or scalelike projections) *Chlamys rubida* (Hinds, 1845)

15b At least many of the radial ribs roughened by spinelike or scalelike projections 16

16a Secondary ribs of left valve much less than half as high as the primary ribs *Chlamys hastata* (Sowerby, 1842)

16b Secondary ribs of left valve about half as high as the primary ribs *Chlamys hastata hericia* (Gould, 1850)
Some consider *C. hericia* to be a species distinct from *C. hastata.* A further complication in our area is a scallop generally called *C. pugetensis* Oldroyd, 1919 (or *C. hastata pugetensis).* It may be only a variant of *C. hastata.* (Grau, 1959, uses the following names: *C. hastata, C. hastata hericia, C. hastata pugetensis.* Bernard (1970) recognizes *C. hastata* and *C. hericia.)*

17a Hinge with taxodont dentition (a series of similar teeth on either side of the umbone of each valve); strictly subtidal 18

17b Hinge without taxodont dentition; includes many intertidal and subtidal species 29

18a Ligament external 19

18b Ligament internal 21

19a Valves with distinct radial sculpture; shell about as high as long (almost circular in outline) *Glycymeris subobsoleta* (Carpenter, 1864)

19b Valves without distinct sculpture; shell decidedly longer than high 20

20a Teeth in the series posterior to the umbone about twice as large as those in the series anterior to it *Malletia faba* Dall, 1897

20b Teeth on both sides of the umbone of about equal size *Tindaria gibbsii* Dall, 1897

21a Interior of valves pearly; without a pallial sinus (animal without siphons) 22

21b Interior of valves not pearly; with a pallial sinus (animal with siphons) 23

22a Valves with a distinct radial sculpture forming chevronlike configurations *Acila castrensis* (Hinds, 1843)

22b Valves essentially smooth *Nucula tenuis* (Montagu, 1808)

23a Shell gaping posteriorly; pallial sinus deep 24

23b Shell closing tightly; pallial sinus shallow 27

24a Shell swollen; posterior end broadly truncate *Yoldia thraciaeformis* Storer, 1838

24b Shell flattened; posterior end not truncate 25

25a Concentric sculpture not coinciding with growth lines

Yoldia scissurata Dall, 1898

25b Concentric sculpture restricted to growth lines 26

26a Umbones slightly anterior to the middle; length more than twice the height *Yoldia limatula* Say, 1831

26b Umbones posterior to the middle; length less than twice the height *Yoldia myalis* Couthouy, 1838

27a Exterior of valves with strong concentric ridges *Nuculana hamata* Carpenter, 1864

27b Exterior of valves without strong concentric ridges 28

28a With 12 teeth anterior to the umbone, 14 posterior to it *Nuculana minuta* Fabricius, 1776

28b With 22 teeth anterior to the umbone, 16 posterior to it *Nuculana cellulita* Dall, 1896

29a Exterior of each valve divided into 2 or 3 regions of distinctly different texture, the anterior region being filelike or rasplike for burrowing into hard clay, rock, or wood; usually with a myophore (a projection for attachment of the foot muscle [not to be confused with a chondrophore, which is a socketlike, spoonlike or shelflike structure to which the internal hinge ligament of some bivalves is attached]); shell gaping anteriorly, where the foot emerges, unless the gape is closed by a callus (as in adult *Penitella)* 30

29b Exterior of valves not usually divided into 2 or 3 regions of distinctly different texture (in a few species, the valves are divided into 2 sectors by a groove originating near the umbones, but the sectors are of similar texture), and without a portion that is filelike or rasplike for burrowing into hard clay, rock, or wood (however, there are a few species in this group that do burrow in clay or rock); without a myophore; shell not often gaping anteriorly, even though it may gape posteriorly 34

30a Shell on the whole almost globular, but appearing to have had a piece notched out of the lower side of the anterior part of each valve; valves without a myophore, but with a strongly developed ridge running from the umbone to the ventral margin; length less than 1 cm; boring in waterlogged wood (subtidal and rare; do not confuse with *Bankia* or *Teredo*, couplet 2) *Xylophaga washingtonia* Bartsch, 1921

30b Shell decidedly elongated, and not appearing to have had a piece notched out of the anterior end (however, the anterior end may be abruptly truncate); with or without a myophore; length regularly exceeding 1 cm; boring in rock or hard clay 31

31a With a well-developed myophore in each valve; posterior portion of shell not tapering sufficiently to resemble a bird's beak (it tapers appreciably, however, in *Pholadidea conradi);* anterior portion of shell gaping anteroventrally, but not appearing to have been abruptly cut off 32

31b Without a myophore in either valve; posterior portion of shell (in mature specimens, at least) elongated and tapering almost to a point, thus resembling a bird's beak; anterior portion very short, gaping as if abruptly cut off *Netastoma rostrata* (Valenciennes, 1845) *N. japonica* (Yokoyama, 1920) has also been reported.

32a Rasping (anterior) portion of each valve occupying about half the area of the valve; valves without a shieldlike plate dorsal to the rasping portion; anteroventral gape not becoming closed, at maturity, by a callus; length up to about 15 cm; boring in hard clay *Zirfaea pilsbryi* Lowe, 1931

32b Rasping portion of each valve approximately triangular, occupying less than half the area of each valve; each valve with a thick, shieldlike plate dorsal to the rasping portion; anteroventral gape becoming closed, at maturity, by a callus; length not often exceeding 7 cm; generally boring in rock, rarely in hard clay 33

33a Posterior (siphonal) ends of valves with thin horny flaps that are continuous with the periostracum; shieldlike dorsal plates equal to half the area of the rasping portions *Penitella penita* (Conrad, 1837)

33b Posterior ends of valves without horny flaps; shieldlike dorsal plates almost equal in area to the rasping portions *Penitella conradi* Valenciennes, 1846

34a Shell as a whole very flat, the thickness not more than about one-sixth the height; right valve actually flat, and fitting into the convex left valve; without a true hinge plate, the right valve with 2 or 3 ridges that articulate with grooves in the left valve 35

34b Shell as a whole at least one-fourth as thick as high; neither valve actually flat; with a true hinge plate, this sometimes provided with teeth or a chondrophore, or both 39

35a Interior of right valve with 3 articulating ridges; dorsal edge of shell, posterior to the umbones, markedly concave *Pandora punctata* Conrad, 1837

35b Interior of right valve with 2 articulating ridges; dorsal edge of shell, posterior to the umbones, nearly straight or slightly convex 36

36a Posterior portion of shell drawn out into a truncate, beaklike prolongation *Pandora filosa* Carpenter, 1864

36b Posterior portion of shell not drawn out into a truncate, beaklike prolongation 37

37a Left valve with a groove that runs from the umbone to the ventral margin, thus dividing the surface into a smaller anterior portion and a larger posterior portion; valves without ridges running from near the umbones to the posterior end 38

37b Left valve without a groove that divides it into 2 portions; valves (especially the left valve) with 2 ridges running from near the umbones to the posterior end *Pandora bilirata* Conrad, 1855

38a Groove on left valve distinct; dorsal margin, behind the umbones, slightly convex *Pandora grandis* Dall, 1877

38b Groove on left valve indistinct; dorsal margin, behind the umbones, straight *Pandora glacialis* Leach, 1819

39a Both valves with a single deep groove that runs from near the umbone to the posterior margin, thus segregating a small posterodorsal sector; hinge without teeth; strictly subtidal 40

39b Neither valve with a deep groove that segregates a small posterodorsal sector; hinge with or without teeth; includes intertidal and subtidal species 42

40a Valves with prominent concentric growth lines; umbones decidedly anterior; length up to about 3 cm *Thyasira bisecta* (Conrad, 1849)

40b Valves almost smooth, the growth lines being very fine; umbones near the middle of the shell; length not over 1 cm 41

41a Umbones prominent; interior of valves glossy, with radiating lines *Thyasira flexuosa* (Montagu, 1803)

41b Umbones not prominent; interior of valves not glossy *Thyasira barbarensis* Dall, 1889

42a With a single adductor muscle scar not far from the center of each valve (height up to about 5 mm, slightly greater than the length; valves prominently beaked, covered externally by a thick periostracum, but beautifully pearly inside; attached by a byssus, usually to coralline algae) *Philobrya setosa* (Carpenter, 1864)

42b With 2 adductor muscle scars in each valve, although the anterior scar may be very small and easily overlooked; height regularly greater than 3 mm, and usually equalled or exceeded by the length 43

43a Mussels--scars of anterior and posterior adductor muscles decidedly unequal (the anterior scar is the smaller one, and may be scarcely noticeable); umbones generally (but not always) much nearer the anterior end than the middle; hinge plate without teeth and without a chondrophore; usually attached to a hard substrate by a byssus of organic threads 44

43b Scars of anterior and posterior adductor muscles almost equal in size (however, they may be different in shape); umbones not generally much nearer the anterior end than the middle (but there are exceptions); hinge plate usually with teeth or a chondrophore, and sometimes with both; not usually attached by a byssus of organic threads (but there are exceptions) 52

44a Shell longer than high; umbones usually at or very close to the anterior end (in *Adula,* near the end of the first quarter) 45

44b Shell about as high or higher than long; umbones near the middle of the dorsal margin 50

45a Umbones essentially terminal; anterior end of valves acute 46

45b Umbones not quite terminal; anterior end of valves rounded 47

46a Valves with a number of coarse radial ribs, as well as irregular growth lines; periostracum usually worn; length frequently exceeding 10 cm; in relatively exposed situations, especially on the outer coast *Mytilus californianus* Conrad, 1837

46b Valves without radial ribs, and with rather regular growth lines; periostracum usually shiny black (brown in very small specimens) and unworn; length rarely exceeding 7 cm; common in protected waters, and not often found on the outer coast (except in bays) *Mytilus edulis* Linnaeus, 1758

47a Umbones about halfway between the anterior end and the middle of the shell; shell almost cylindrical, with dorsal and ventral margins nearly parallel for at least half the length; boring in rock *Adula californiensis* (Philippi, 1847)

47b Umbones closer to the anterior end than to the middle of the shell; shell not almost cylindrical, the dorsal and ventral margins not parallel 48

48a Length at least twice the height; periostracum brown, in young specimens elaborated into soft bristles *Modiolus rectus* (Conrad, 1837)

48b Length from about 1 and three-fourth times to almost twice the height; periostracum generally blackish brown, not elaborated into bristles 49

49a Length about 1 and three-fourth times the height *Musculus laevigatus* (Gray, 1824)

49b Length almost twice the height *Musculus niger* (Gray, 1824)

50a Shell with radial ribs (these may be fine); subtidal, but not restricted to very deep water 51

50b Shell smooth, without any radial ribs; restricted to very deep water (at least 500 m) *Dacrydium pacificum* Dall, 1916

51a Radial ribs fine, inconspicuous *Solamen columbianum* (Dall, 1897)

51b Radial ribs prominent *Crenella decussata* (Montagu, 1808)

52a Hinge plate of both valves without teeth (except in very young *Hiatella)* and without a chondrophore 53

52b Hinge plate of 1 or both valves either with teeth or with a chondrophore, or both (the teeth are weakly developed in *Serripes*, which has a heavy shell with low radial ribs and thus resembles a cockle, and which is unlike any of the species covered in choice 52a) 61

53a Valves generally nearly circular in outline, but not consistently the same shape, and often obviously deformed; embedded in colonies of compound ascidians, with the siphonal end near the surface *Mytilimeria nuttallii* Conrad, 1837

53b Valves decidedly longer than high, not nearly circular in outline (often deformed in *Lyonsia* and *Hiatella);* not embedded in colonies of compound ascidians (but may be embedded in sponge masses or nestling in burrows made by other animals) 54

54a Umbones so close together that they touch, the left one usually wearing a depression or hole in the right one; strictly subtidal 55

54b Umbones not so close together that they touch; intertidal and subtidal 58

55a Anterior and posterior ends of each valve almost identically rounded; umbones almsot exactly in the middle *Thracia beringi* Dall, 1915

55b Anterior and posterior ends of each valve dissimilar; umbones slightly nearer the anterior end than the posterior end 56

56a Umbones about halfway between the end of the first third of the shell and the middle; length up to about 5 cm 57

56b Umbones near the end of the first third of the shell; length not often exceeding 3 cm *Thracia curta* Conrad, 1837

57a Posterior end of the shell nearly truncate *Thracia trapezoides* Conrad, 1849

57b Posterior end of the shell broadly rounded, not truncate *Thracia challisiana* Dall, 1915

58a Interior of valves usually at least slightly pearly; with a continuous pallial line and pallial sinus; siphons without red tips 59

58b Interior of valves chalky white, not at all pearly; pallial line discontinuous, pallial sinus absent; siphons with bright red tips *Hiatella* spp.

Two species, *H. arctica* (Linnaeus, 1767) and *H. pholadis* (Linnaeus, 1767), are reported. They are difficult to differentiate. *H. arctica* is probably the more common. It can burrow into soft rock, but is often a nestler in burrows of pholads, in algal holdfasts, and other protected situations. *H. pholadis* seems to be strictly a nestler in burrows of pholads; it is larger (up to about 5 cm) and its umbones are very close to the anterior end.

59a Shell not often appearing to be deformed; valves with delicate radial ribs, little periostracum (often pearly externally), and not so thin as to usually crack when dried *Lyonsia (Lyonsia) californica* Conrad, 1837

59b Shell usually at least somewhat deformed; valves without radial ribs, with abundant periostracum, and so thin as to usually crack when dried 60

60a Valves gaping posteriorly or ventrally; length occasionally exceeding 10 cm; not living in sponge masses *Lyonsia (Entodesma) saxicola* Baird, 1863

60b Valves not gaping; length up to about 2.5 cm; living in sponge masses

Lyonsia (Entodesma) inflata Conrad, 1837

61a Shell more than twice as long as high; periostracum glossy, yellowish olive or greenish in color 62

61b Shell rarely twice as long as high (and usually not more than 1 and one-half times as long as wide); periostracum sometimes glossy and sometimes greenish, but not in species that come close to being twice as long as wide 63

62a Shell about 2 and one-fourth times as long as high; restricted to exposed sandy beaches of the open coast *Siliqua patula* (Dixon, 1789)

62b Shell about 4 times as long as wide; usually in flats of slightly muddy sand (subtidal as well as intertidal) *Solen sicarius* Gould, 1850

63a Anterior two-thirds of shell almost globular and radially ribbed, the posterior third abruptly narrowed into a smoother, almost cylindrical, truncate beak; length rarely exceeding 1 cm 64

63b Not distinctly divided into an almost globular anterior portion and an abruptly narrower truncate beak (valves, if radially ribbed, with ribs usually distributed over almost all areas of the surface); length frequently greater than 1 cm 65

64a Height slightly more than half the length; with about 15 radial ribs *Cardiomya oldroydi* (Dall in Oldroyd, 1924)

64b Height about half the length; with 16 to 20 radial ribs *Cardiomya californica* (Dall, 1924)

These are the only septibranchs regularly encountered in our area. The distinctions between them need to be sharpened. Both species are epibenthic on muddy bottoms, occurring at depth of about 20 m and greater. Some other species have been reported from this general region, but at least certain of these seem to be restricted to very deep water.

65a Valves with distinct radial ribs (these may be weakly developed or very delicate, but any bivalves with radial sculpture are in this section) 66

65b Valves without radial ribs 80

66a Length of shell not appreciably greater (and sometimes slightly less than) the height 67

66b Length of shell at least 1 and one-fourth times the height 76

67a Posterior quarter of each valve with rather prominent concentric ridges and very low radial ribs, so that the sculpture in this area is cancellate and decidedly different from that over the rest of the surface, where only the closely-spaced radial ribs are distinct *Nemocardium centifilosum* (Carpenter, 1864)

67b Valves without a considerable area in which the sculpture is different from that of the rest of the surface 68

68a Radial ribs very low or delicate, generally overshadowed by the concentric ridges 69

68b Radial ribs prominent, not overshadowed by the concentric ridges 70

69a Without a pallial sinus; exterior of valves white, with a grayish or greenish periostracum; interior without a purple stain at each side of the hinge; length up to about 1 cm; subtidal *Lucinoma tenuisculpta* Carpenter, 1864

69b With a pallial sinus; exterior of valves with pink radial rays, and usually with some greenish periostracum; interior of valves with a purple stain at each end of the hinge; length up to about 3 cm; intertidal and subtidal *Semele rubropicta* Dall, 1871

70a Hinge with both cardinal teeth and lateral teeth; with more than 30

radial ribs; height regularly more than 2 cm 71
70b Hinge with only cardinal teeth; with fewer than 30 radial ribs; height generally under 2 cm 73
71a Shell as high as long; height up to about 8 cm; common intertidal species *Clinocardium nuttallii* (Conrad, 1837)
71b Shell slightly longer than high; height rarely exceeding 4 cm; subtidal 72
72a With about 45 to 50 ribs (sometimes more) *Clinocardium californiense* (Deshayes, 1841)
72b With about 40 ribs *Clinocardium ciliatum* (Fabricius, 1780)
73a With not more than 15 ribs 74
73b With more than 15 ribs 75
74a Height about equal to the length; with 12 to 14 ribs *Cardita paucicostata* Krause, 1885
74b Height slightly greater than the length; with 10 to 11 ribs *Miodontiscus prolongata* Carpenter, 1864
75a With about 19 to 20 ribs *Cardita ventricosa* Gould, 1850
75b With about 25 ribs *Cardita crebricostata* Krause, 1885
76a Unbomes near the middle of the shell; radial ribs very weak, though rather broad, and the surface of the valves almost smooth *Serripes groenlandicus* Bruguière, 1789
76b Umbones decidedly anterior to the middle of the shell; radial ribs usually distinct, even if delicate 77
77a With about 14 to 15 radial ribs; anteroventral corner of each valve forming nearly a right angle; height up to about 1 cm *Glans subquadrata* (Carpenter, 1864)
77b With many more than 15 radial ribs; anteroventral corner of each valve not forming nearly a right angle; height usually more than 1 cm 78
78a Radial ribs typically more prominent than the concentric ridges; valves often with zigzag or chevronlike markings, especially in younger specimens; hinge ligament about one-third the length of the shell; length not exceeding 6 cm 79
78b Radial ribs less prominent than the concentric ridges; valves without zigzag or chevronlike markings; hinge ligament nearly half the length of the shell; length sometimes exceeding 8 cm *Protothaca tenerrima* (Carpenter in Gould & Carpenter, 1857)
79a With a series of toothlike elevations on the inside of the valves close to the margins; interior of valves whitish, without a purple stain near the posterior end *Protothaca staminea* (Conrad, 1837)
79b Without toothlike elevations on the inside of the valves close to the margins; interior of valves yellowish, with a purple stain near the posterior end *Tapes japonica* (Deshayes, 1853)
80a With a shelflike or spoonlike chondrophore projecting from the hinge plate of 1 valve; hinge plate of both valves without teeth; hinge ligament largely or completely internal 81
80b Either without a chondrophore in either valve, or with a socketlike chondrophore in both valves; hinge plates of valves with teeth; hinge ligament internal or external 85
81a Pallial sinus at least as deep as high; umbones at least slightly anterior to the middle; anterior and posterior halves of valves dissimilar 82
81b Pallial sinus very shallow; umbones almost exactly in the middle; anterior and posterior halves of shell similar (siphons opening into burrows of *Upogebia* and *Callianassa*)

Cryptomya californica (Conrad, 1837)

82a Chondrophore not projecting very far, elongated in the longitudinal axis of the shell; umbones very decidedly anterior to the middle (anterior and posterior portions of the valves unlike, the posterior portion narrower and almost truncate; in burrows of other animals and in algal holdfasts) *Sphenia ovoidea* Carpenter, 1864

82b Chondrophore projecting nearly or quite to the extent of its width; umbones only slightly anterior to the middle 83

83a Posterior (siphonal) end of shell somewhat less smoothly rounded than the anterior end, but not truncate; length sometimes exceeding 10 cm; especially common in estuarine situations where the salinity is reduced *Mya arenaria* Linnaeus, 1758

83b Posterior end of shell almost truncate; length rarely exceeding 7 cm; nor normally in estuarine situations 84

84a Shell about 1 and one-half times as long as high; periostracum thick, brown, wrinkled; interior of valves yellowish, not pearly *Mya truncata* Linnaeus, 1758

84b Shell almost twice as long as high; periostracum thin, yellowish; interior of valves slightly pearly *Platyodon cancellatus* (Conrad, 1837)

85a Valves without a pallial sinus 86

85b Valves with a pallial sinus 97

86a Valves with strong concentric ridges, and not nearly disk-shaped 87

86b Valves without strong concentric ridges (only rather fine growth lines) except in 1 species *(Lucinoma annulata)* which has nearly disk-shaped valves 89

87a Radial ridges irregular (some not complete, others splitting into 2) *Astarte esquimalti* Baird, 1863

87b Radial ridges rather regular 88

88a Length sometimes exceeding 2.5 cm; umbones distinctly anterior to the middle *Astarte alaskensis* Dall, 1903

88b Length up to about 1.2 cm; umbones almost in the middle *Astarte compacta* Carpenter, 1864

89a Height approximately equal to the length, the valves almost disk-shaped 90

89b Length decidedly greater than the height, the valves not disk-shaped 92

90a With some conspicuously raised concentric ridges, the shell thus noticeably roughened; length up to about 5 cm *Lucinoma annulata* (Reeve, 1850)

90b Without raised concentric ridges (only relatively fine growth lines); length not exceeding 2 cm 91

91a Exterior of valves pale green, glistening when fresh; length up to a-about 7 mm *Axinopsida serricata* (Carpenter, 1864)

91b Exterior of valves whitish, not glistening; length up to about 2 cm, usually smaller *Diplodonta orbellus* (Gould, 1852)

92a Ventral margin of valves almost straight or even slightly indented; up to about 2 cm long; attached by a byssus to the underside of the mud shrimp, *Upogebia*, and the sea mouse, *Aphrodita* (rarely found temporarily attached to other hosts) *Orobitella rugifera* (Carpenter, 1864)

92b Ventral margins of valves decidedly convex; up to about 2.5 cm long *(Kellia)*, but generally less than 5 mm long; not attached to *Upogebia* or *Aphrodita*, but frequently nestling in burrows, in algal holdfasts, or in beds of barnacles and mussels 93

93a Periostracum greenish or olive-yellow; mantle capable of being extended to cover the whole shell; inhalant siphon directed anteriorly, the exhalant siphon directed posteriorly; up to about 2.5 cm long 94

93b Periostracum not greenish or olive-yellow; mantle not capable of being extended to cover a large part of the shell; both siphons directed posteriorly; length not over 5 mm 95

94a Shell slightly more than 1 and one-fourth times as long as high *Kellia laperousii* Deshayes, 1839

94b Shell not quite 1 and one-fourth times as long as high *Kellia suborbicularis* (Montagu, 1804)

95a Umbones slightly closer to the posterior end than to the middle; with abundant dark brown periostracum; interior of valves slightly pearly *Mysella tumida* Carpenter, 1864
M. aleutica Dall, 1899 and *M. compressa* (Dall, 1913) have also been reported.

95b Umbones about halfway between the anterior end and the middle, or slightly closer to the middle than this; without a dark brown periostracum; interior of valves not pearly 96

96a Exterior of valves purplish in the vicinity of the umbones, interior white *Turtonia minuta* (Fabricius, 1780)

96b Exterior and interior of valves reddish *Lasaea cistula* Keen, 1938

97a Pallial line consisting of a series of disconnected scars (with 1 small tooth on the hinge plate of each valve; gaping at both ends; posterior end truncate; up to about 7 cm long; subtidal) *Panomya ampla* Dall, 1898

97b Pallial line continuous 98

98a With a single tooth on the hinge plate of each valve; shell very large, sometimes exceeding a length of 15 cm, but not large enough to accommodate the immense siphon, and gaping widely except in the region of the hinge *Panopea generosa* (Gould, 1850)

98b With more than 1 tooth on the hinge plate of each valve; shell generally less than 10 cm long (except in *Tresus),* large enough to accommodate the siphon when it is withdrawn, gaping conspicuously only at the posterior end, if at all 99

99a Shell gaping broadly at the posterior end (the width of the gape is more than one-fourth the width of the shell); length regularly exceeding 12 cm 100

99b Shell closing tightly, or gaping only slightly at the posterior end; length rarely exceeding 10 cm 101

100a Shell generally about 1 and one-half times as long as high; umbones near the end of the anterior third; common and widely distributed in our region *Tresus capax* (Gould, 1850)

100b Shell more than 1 and one-half times as long as wide; umbones near the end of the anterior quarter; scarce in our region *Tresus nuttallii* (Conrad, 1853)

101a Hinge ligament largely internal 102

101b Hinge ligament external 103

102a Internal portion of the ligament separated from the external portion by a calcareous plate *Mactra californica* Conrad, 1837

102b Internal and external portions of the ligament somewhat distinct, but not separated by a calcareous plate *Spisula falcata* (Gould, 1850)

103a With 3 cardinal teeth (and sometimes also a pair of lateral teeth) in each valve 104

103b With fewer than 3 cardinal teeth in at least 1 of the valves 110

104a Length less than 1 cm 105

104b Length regularly more than 2 cm 107

105a Umbones almost perfectly central, the outline of the valves resembling a rounded isosceles triangle; inside of ventral margins of valves smooth *Transennella tantilla* (Gould, 1853)

105b Umbones at least slightly anterior to the middle; inside of ventral margins of valves smooth or crenulated 106

106a Inside of ventral margins of valves smooth; exterior without a purplish tinge; pallial sinus directed upward at an angle of about 45° *Psephidia lordi* (Baird, 1863)

106b Inside of ventral margins of valves crenulated; exterior often with a purplish tinge; pallial sinus directed upward at an angle of about 60° (locally introduced from the Atlantic coast) *Gemma gemma* (Totten, 1834)

107a With an elongated lateral tooth on each side of the 3 cardinal teeth; shell gaping slightly at the posterior end; concentric ridges raised, but not also bent in the direction of the umbones; length sometimes exceeding 10 cm; common intertidally *Saxidomus giganteus* Deshayes, 1839

107b Without a lateral tooth on each side of the 3 cardinal teeth; shell closing tightly at the posterior end; concentric ridges, if raised significantly, also bent in the direction of the umbones; length variable (maximum ranging from about 2 cm to about 10 cm); subtidal 108

108a Outline of valves somewhat triangular, the umbones not far from the middle; with a polished yellow-brown periostracum; length up to about 2 cm *Liocyma scammoni* Dall, 1871

108b Outline of valves more nearly oval than triangular, the umbones near the end of the first one-third; without a polished periostracum, and usually whitish; length regularly exceeding 5 cm 109

109a Valves with some of the concentric ridges conspicuously raised and generally also bent in the direction of the umbones; length up to about 10 cm *Humilaria kennerlyi* (Reeve, 1863)

109b Valves without conspicuously raised ridges; length up to about 6 cm *Compsomyax subdiaphana* (Carpenter, 1864)

110a With 2 cardinal teeth in each valve 112

110b With 3 cardinal teeth in one valve, 2 in the other 111

111a Valves oval, rather delicate, polished; siphons not tipped with bright purple *Cooperella subdiaphana* (Carpenter, 1864)

111b Valves somewhat elongated (usually narrower posteriorly than anteriorly), rather thick, chalky (often distorted as a result of conforming to the outline of pholad burrows in which this species nestles); siphons tipped with bright purple *Petricola carditoides* (Conrad, 1837)

112a Valves almost ellipsoidal, about twice as long as high; hinge plate of each valve, behind the umbone, elaborated into a thin dorsal extension to which the ligament is attached; exterior of valves usually with pinkish or purplish rays; siphons short, united; length up to about 7 cm *Gari californica* (Conrad, 1837)

112b Valves not almost ellipsoidal (one end is usually different from the other, but if not, the ends are not evenly rounded); hinge plate not elaborated into a thin dorsal extension; exterior of valves without pinkish or purplish rays (however, various tints may be present internally or externally); siphons long, separate; maximum length ranging from about 1 cm to about 10 cm 113

113a Hinge plate with lateral teeth (these may be small) in addition to the 2 cardinal teeth 114

113b Hinge plate without lateral teeth in addition to the 2 cardinal teeth 117

114a Shell about twice as long as high; length regularly more than 1 cm (maximum about 2 cm or about 5 cm) 115

114b Shell decidedly less than twice as long as high; length not much more than 1 cm 116

115a Interior of each valve with a ridge just behind the scar of the anterior adductor muscle; length not often exceeding 2 cm *Tellina modesta* (Carpenter, 1864)

T. buttoni Dall, 1900 is very similar, if not identical.

115b Interior of valves without a ridge behind the scar of the anterior adductor muscle; length up to about 5 cm *Tellina bodegensis* Hinds, 1845

116a Shell slightly more than 1 and one-half times as long as high; valves becoming narrowed toward the posteroventral corner; exterior white to pink; length up to slightly more than 1 cm *Tellina carpenteri* Dall, 1900

116b Shell about 1 and one-third times as long as high; valves rounded posteriorly; exterior whitish or yellow, but interior often pink *Tellina nuculoides* (Reeve, 1854)

117 The genus *Macoma* is represented in this region by about 12 species. The key offered here covers only the 4 species likely to be found intertidally. Some of the subtidal species are difficult to segregate from these. Detailed reviews of the genus, as represented in this area, have been published by Dunnill & Ellis (1969), and Coan (1971), and anyone involved with the subtidal fauna is referred to these papers.

117a Posterior portions of valves distinctly bent to the right *Macoma nasuta* Conrad, 1837

117b Posterior portions of valves not bent to the right 118

118a Periostracum highly polished, usually transparent; hinge ligament not more than one-fourth the length of the slope from the umbones to the posterior end; length regularly exceeding 5 cm, and sometimes attaining 10 cm *Macoma secta* (Conrad, 1837)

118b Periostracum not highly polished and transparent; hinge ligament at least one-third the length of the slope from the umbones to the posterior end; length not more than about 5 cm 119

119a Outline of valves almost regularly oval, the posterior end nearly as rounded as the anterior end; interior of valves often yellow, pink, or faintly bluish; length up to about 2 cm *Macoma balthica* (Linnaeus, 1758)

119b Outline of valves not almost regularly oval, the posterior end more narrow and less rounded than the anterior end; interior of valves white; length up to about 5 cm *Macoma inquinata* (Deshayes, 1855)

Coan, E. V. 1971. The northwest American Tellinidae. Veliger, 14 (suppl.): 1-63.

Dunnill, R. M., & D. V. Ellis. 1969. Recent species of the genus *Macoma* (Pelecypoda) in British Columbia. Nat. Mus. Canada, Nat. Hist. Pap., 45:1-34.

Fitch, J. E. 1953. Common marine bivalves of California. Calif. Dept. Fish & Game, Fish. Bull. no. 90. 102 pp.

Grau, G. 1959. Pectinidae of the eastern Pacific. Allan Hancock Found. Pa-

cific Exped., 23:i-vii, 1-308.
Pearce, J. B. 1965. On the distribution of *Tresus capax* and *Tresus nuttallii* in the waters of Puget Sound and the San Juan Archipelago (Pelecypoda: Mactridae). Veliger, 7:166-71.
Quayle, D. B. 1969. Intertidal Bivalves of British Columbia. Handbook 17, British Columbia Provincial Museum, Victoria. 2nd ed. 104 pp.
Soot-Ryen, T. 1955. A report on the family Mytilidae (Pelecypoda). Allan Hancock Found. Pacific Exped., 20:1-174.
Turner, R. D. 1966. A survey and Illustrated Catalogue of the Teredinidae (Mollusca: Bivalvia). Cambridge, Massachusetts: Museum of Comparative Zoology, Harvard University. vii + 265 pp.

CLASS SCAPHOPODA

Although there are only a few scaphopods in our area, these are poorly known. Oldroyd (1924) provided brief descriptions (taken verbatim from the original sources) of the various species reported from this region, and subsequently (1927) did the same thing for all scaphopods recorded from the Pacific coast. In most cases, it is not possible to decide which species of each of the 2 genera, *Cadulus* and *Dentalium*, are valid, and to what extent they differ from one another. Bernard (1970) reported 3 species of *Cadulus* and 3 of *Dentalium* from British Columbia waters. These are listed below, in connection with the key to genera.

1a Shell widest at the aperture; length regularly more than 2 cm, and sometimes up to about 6 cm ... *Dentalium*
Species reported by Bernard (1970): *D. agassizii* Pilsbry & Sharp, 1897; *D. pretiosum* Sowerby, 1860; *D. rectius* Carpenter, 1864.

1b Shell widest at a level posterior to the aperture; length generally less than 1.5 cm ... *Cadulus*
Species reported by Bernard (1970): *C. aberrans* Whiteaves, 1887; *C. fusiformis* Pilsbry & Sharp, 1897 (=*C. hepburni* Dall, 1897); *C. tolmiei* Dall, 1897.

Clark, R. B. 1963. The economics of *Dentalium*. Veliger, 6:9-19.
Emerson, W. K. 1962. A classification of the scaphopod mollusks. J. Paleontol., 36:461-82.
Pilsbry, H. A., & B. Sharp. 1897-98. Class Scaphopoda. *In* Manual of Conchology, 17:i-xxxii, 1-280.

CLASS CEPHALOPODA

F. G. Hochberg
Santa Barbara Museum of Natural History

The key includes pelagic species that are likely to be found near the surface close to shore, and benthic species that occur in moderately shallow water. The papers of Bernard (1970), Berry (1912), Pearcy (1965), and Young (1972) list species that may be found near the surface far from shore or on the bottom in deep water, or that may be collected by mid-water trawls. Young's report has a substantially complete bibliography on Pacific coast cephalopods.

1a With 8 arms; arms with sessile suckers; body ovoid and without fins; without an internal shell or "pen" 2

1b With 2 tentacles in addition to 8 arms; arms with stalked suckers and/or hooks; body rounded or elongate, with fins; with an internal shell or "pen" 4

2a Dorsal mantle length (measured from the apex to a point midway between the eyes) frequently exceeding 20 cm; body with extensive skin folds; length of arms 3 to 5 times the length of the body; with 24 to 29 primary gill lamellae on each gill
Octopus dofleini (Wülker, 1910), var. *martini* Pickford, 1964

2b Dorsal mentle length less than 6 cm; body either without skin folds or with only 1 distinct fold on each side; length of arms not more than about 4 times the length of the body; with 18 to 23 gill lamellae on each gill 3

3a Body smooth, not papillose; with a distinct fold (much like a keel) bordering the body laterally; length of arms 2 to 3 times the length of the body; hectocotylus of male about one-fifth the length of the arm of which it is a part; without enlarged suckers
Octopus leioderma Berry, 1911

3b Body without a skin fold, but papillose or cirrose; length of arms 3 to 4 times the length of the body; hectocotylus of male about one-tenth the length of the arm of which it is a part; enlarged suckers present on all but the ventral arms of the adults of both sexes
Octopus sp. (undescribed)

4a Body rounded; dorsal mantle length (measured from the apex to the free edge of the mantle) less than 4 cm; mantle free all around; fins semicircular, almost as long as the mantle, with a broad free lobe; arms short, circular in cross-section; tentacles retractile into special pits; dorsal arms of male hectocotylized *Rossia pacifica* Berry, 1911

4b Body elongate and tapering; dorsal mantle length usually greater than 5 cm; mantle attached; fins triangular, generally less than half as long as the mantle; arms long, angular in cross-section; tentacles not retractile into special pits; hectocotylized arms present or absent in male 5

5a Arms of lowermost pair much elongated (in the male, the left arm of this pair is hectocotylized); without hooks on the tentacles or on the arms; tentacles without modified suckers that form a "fixing apparatus;" eyes covered by a continuous membrane (cornea); dorsal mantle length not exceeding 15 cm *Loligo opalescens* Berry, 1911

5b Arms of lowermost pair not especially elongated (in the male, the left arm of this pair is not hectocotylized); with hooks on either the tentacles or on some of the arms (the hooks are highly modified suckers); tentacles with modified suckers that form a "fixing apparatus;" membrane over the eyes perforated; dorsal mantle length may exceed 20 cm 6

6a Arms (except those of the lowermost pair) with hooks as well as suckers; fins less than half the mantle length; dorsal mantle length up to about 30 cm (specimens collected from floats are usually small, having a mantle length of only about 10 cm or less)
Gonatus ?fabricii (Lichtenstein, 1818)

6b Arms without hooks (hooks are present on the tentacle clubs, however); fins more than half the mantle length; dorsal mantle length commonly exceeding 1.5 m *Moroteuthis robusta* (Verrill, 1876)

Bernard, F. R. 1970. A distributional checklist of the marine molluscs of British Columbia: based on faunistic surveys since 1950. Syesis, 3:75-94.

Berry, S. S. 1911. Preliminary notices of some new Pacific cephalopods. Proc. U. S. Nat. Mus., 37:407-19.

------. 1912. A review of the cephalopods of western North America. Bull. U. S. Bur. Fisheries, 30 (1910):267-336.

Fields, W. G. 1965. The structure, development, food relations, reproduction, and life history of the squid *Loligo opalescens* Berry. Calif. Dept. Fish & Game, Fish. Bull. no. 131. 108 pp.

Fields, W. G., & Gauley, V. A. 1971. Preliminary descriptions of an unusual gonatid squid (Cephalopoda: Oegopsida) from the north Pacific. Jour. Fish. Res. Bd. Canada, 28:1796-1801.

Pattie, B. H. 1968. Notes on giant squid *Moroteuthis robusta* (Dall) Verrill trawled off the southwest coast of Vancouver Island, Canada. Wash. Dept. Fish., Res. Rep., 3:47-50.

Pearcy, W. G. 1965. Species composition and distribution of pelagic cephalopods from the Pacific Ocean off Oregon. Pacific Sci., 19:261-6.

Pearcy, W. G., & Voss, G. L. 1963. A new species of gonatid squid from the northeastern Pacific. Proc. Biol. Soc. Wash., 76:105-12.

Pickford, G. E. 1964. *Octopus dofleini* (Wülker), the giant octopus of the north Pacific. Bull. Bingham Oceanogr. Coll., 19 (1):1-68.

Young., R. E. 1972. The systematic and areal distribution of pelagic cephalopods from the seas off Southern California, Smithsonian Contr. Zool., no. 97. iii + 159 pp.

PHYLUM ANNELIDA

CLASS POLYCHAETA

The polychaete fauna of our region is extremely rich and varied. It is not feasible to include, in a manual of this kind, all species that have been recorded or that may be expected. Some of them are relatively rare; others are small and easily overlooked unless samples of subtidal mud are carefully sorted. The obvious advantage of having practical keys for identification of most of the intertidal, subtidal, and planktonic polychaetes likely to be encountered in our area will offset the disadvantage of their being incomplete. Moreover, there are special publications--some of them comprehensive--that will enable one to cope with almost any polychaete once it has been identified at least to family. The bibliography lists a number of especially valuable treatises concerned with our fauna, and also a few papers of more restricted scope that are useful in dealing with certain families or genera, or with the polychaetes found in plankton.

For over 30 years, biologists of this region have relied heavily on the 2-part study of Berkeley & Berkeley (1948, 1952). Banse and Hobson (1974) have prepared a modernized and greatly expanded version of Berkeley and Berkeley's treatment of "errant" polychaetes, including almost every species that has been reported or that might be expected. Their work also includes an up-to-date bibliography. A similar treatise on "sedentary" polychaetes is in preparation.

Attention is also called to the 2 atlases of California polychaetes prepared by Hartman (1968, 1969). These are admirably illustrated, and will be helpful in recognizing most genera represented in our region, and frequently the species as well. Numerous other papers by Hartman, many of them concerned with systematics within individual families, are not cited here. *Light's Manual* and Banse and Hobson (1974) have detailed bibliographies covering her work, as well as that of many other authors concerned with Pacific coast polychaetes. Hartman (1951) provides a substantially complete guide to the literature on polychaetes, on a worldwide basis, up to about 1950.

For convenience, the treatment of polychaetes in this manual is divided into 2 sections: a key to families, and an alphabetical listing of the families, with further keys to the species.

Key to Families

1a Body short, grublike, with a dark, kidney-shaped horny plate and a tuft of filamentous gills at the posterior end — Sternaspidae

1b Body generally not grublike, and without a dark, horny plate and filamentous gills at the posterior end — 2

2a Body almsot perfectly transparent; with a pair of long, tentaclelike parapodia (supported by fine setae) on each side of the head, in addition to about 20 pairs of swimming parapodia in which 2 lobes arise from the end of a rather long stalk (length generally under 2 cm; strictly planktonic) — Tomopteridae

2b Body not generally transparent; without a pair of long, tentaclelike parapodia on each side of the head, and if the parapodia are modified for swimming, the lobes generally arise close to the body, not at the end of a stalk — 3

3a Head (i.e., the region represented by the prostomium and the first 1 or 2 segments) without antennae, palps, tentacular cirri, or branched filaments, and without setae that cover it or that function as part of a device for closing a tube in which the animal lives (however, in the Oweniidae, the anterior end has a membrane with frilly lobes) — 4

3b Head with antennae, palps, tentacular cirri, or branched filaments, or with setae that either cover much of the head or that function as part of a device for closing a tube in which the animal lives (in the Euphrosinidae, which are rare, the antennae are small, and visible only if the prostomium is viewed from the ventral side) — 13

4a Prostomium indistinct; surface of the body with scattered papillae, many of which are almost globular (subtidal and rare) — Sphaerodoridae

4b Prostomium distinct; papillae, if present in the body surface, not almost globular — 5

5a Most of the segments of at least the anterior two-thirds of the body decidedly longer than wide — 6

5b Few if any of the segments in the anterior two-thirds of the body longer than wide — 7

6a Head with a frilly lobed membrane; posterior end of the body not in the form of a funnellike structure — Oweniidae

6b Head without a frilly lobed membrane, but with an oblique, nearly flat platelike area; posterior end usually in the form of a funnellike structure (this is often fringed) — Maldanidae

7a With several pairs of large, bushy gills in the middle third of the body — Arenicolidae

7b Without several pairs of large, bushy gills in the middle third of the body (but small, generally simple gills may be present in this region, and conspicuous bushy gills may be present near the anterior end — 8

8a All setae of a single type: very slender bristles Opheliidae
8b Setae of more than 1 type, which may include slender bristles, forked bristles, stout hooks, hooks with hoodlike elaborations, etc. 9
9a Proboscis with dark, hard jaw pieces (to see these, it will be necessary to cause the proboscis to be everted by applying gentle pressure behind the head); dorsal gills absent 10
9b Proboscis without hard jaw pieces; dorsal gills present or absent 11
10a Neurosetae including hooded hooks; eyes absent Lumbrineridae
10b Neurosetae not including hooded hooks; eyes sometimes present Arabellidae
11a Proboscis simple, sometimes papillate, but not in the form of a rosette around the mouth and not consisting of branched lobes 12
11b Proboscis either in the form of a rosette around the mouth or consisting of branched lobes Orbiniidae
12a Body as a whole very slender and threadlike; setae consisting of slender bristles and hooded hooks, but not of forked bristles; prostomium conical; without bushy dorsal gills on a few segments near the anterior end (however, small gills are usually present over a rather large part of the body, beginning about 10 or 15 segments from the anterior end) Capitellidae
12b Body not slender and threadlike, usually somewhat swollen in the anterior half; setae consisting of slender bristles and forked bristles, but not of hooded hooks; prostomium usually bilobed or drawn out into 2 anterolateral "horns;" bushy dorsal gills generally present on a few segments close to the anterior end Scalibregmidae
13a Anterior end with a crown of several to many featherlike filaments, the complex formed by these filaments resembling a feather-duster 14
13b Anterior end without a crown of featherlike filaments (however, there may be antennae, palps, or tentacular cirri on the head) 15
14a Secreting a calcareous tube closed by a funnel-shaped or globular operculum Serpulidae
14b Secreting a leathery, parchmentlike, or mucous tube, not closed by an operculum Sabellidae
15a Much or all of the dorsal surface covered by scalelike elytra, feltlike bristles, or paleae (fanlike aggregates of flattened setae) 16
15b Dorsal surface not covered by scalelike elytra, feltlike bristles, or paleae 20
16a Much of dorsal surface covered by scalelike elytra 17
16b Dorsal surface without elytra, but covered by feltlike bristles or fanlike aggregrates of flattened setae 19
17a At least some of the neurosetae compound (consisting of 2 separate portions) 18
17b All of the setae simple Polynoidae
18a Elytra and dorsal cirri alternating regularly from the fifth setigerous segment to the posterior end Polyodontidae
18b Elytra and dorsal cirri alternating only in the anterior part of the body (in the posterior part, elytra occur on all segments, and there are no dorsal cirri) Sigalionidae
19a Dorsal surface covered by feltlike bristles; length regularly exceeding 3 cm Aphroditidae
19b Dorsal surface covered by fanlike aggregates of flattened setae; length generally not exceeding 1 cm Chrysopetalidae
20a Notosetae on each side of the body forming transverse rows that reach nearly to the midline; gills (up to several pairs on each side of some

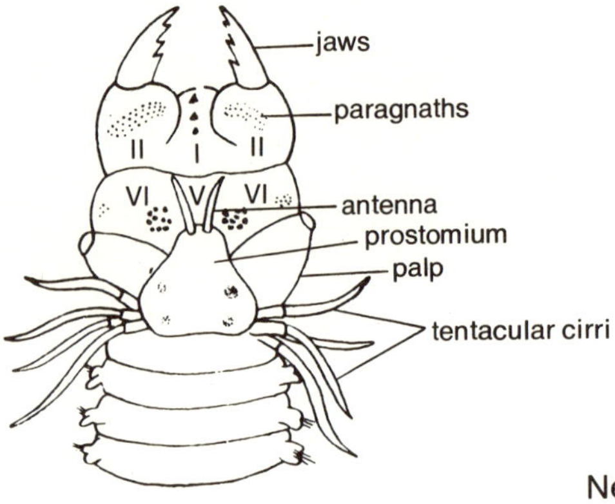

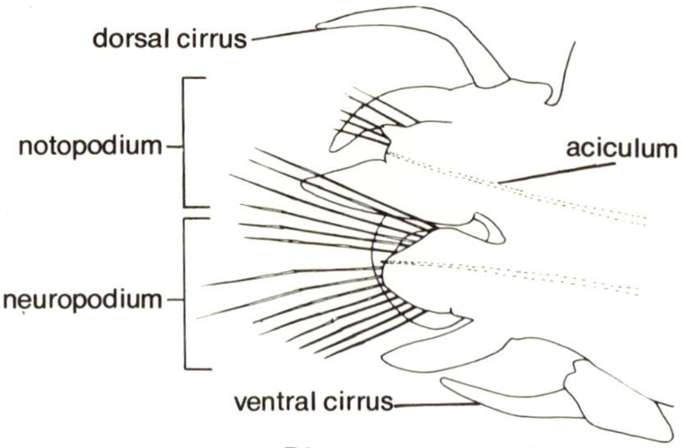

Biramous parapodium

Nereidae

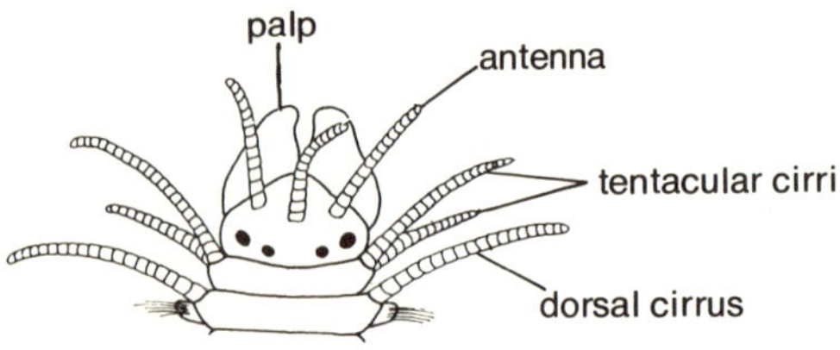

Syllidae

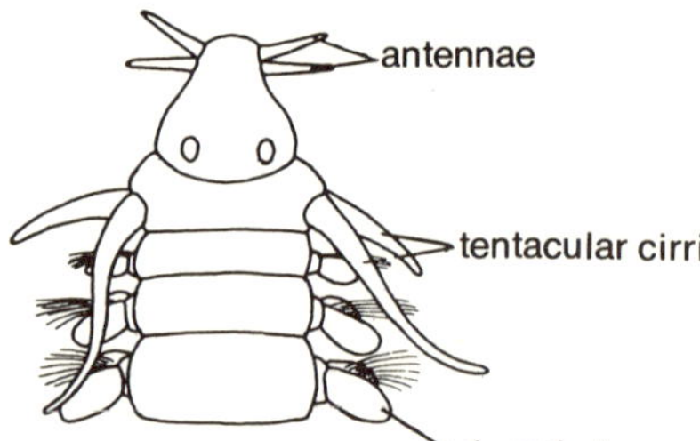

Phyllodocidae

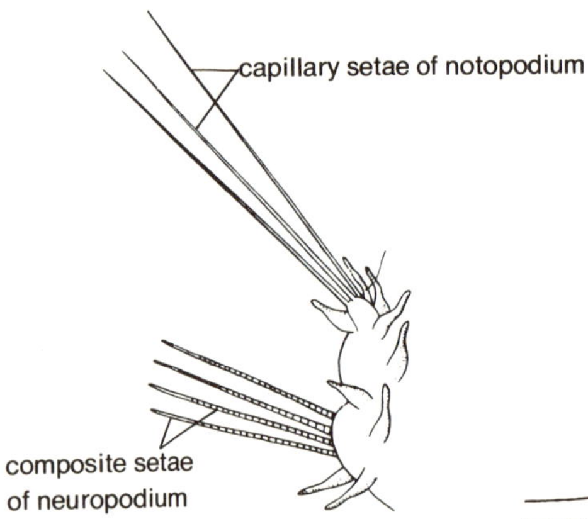

Biramous parapodium

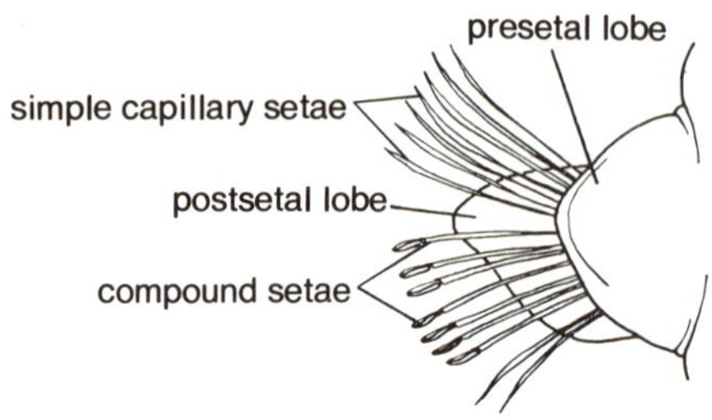

Uniramous parapodium
(anterior aspect)

Hooded hook

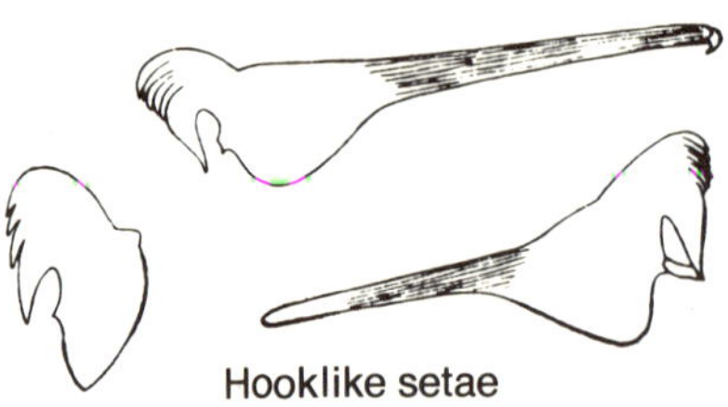

Hooklike setae

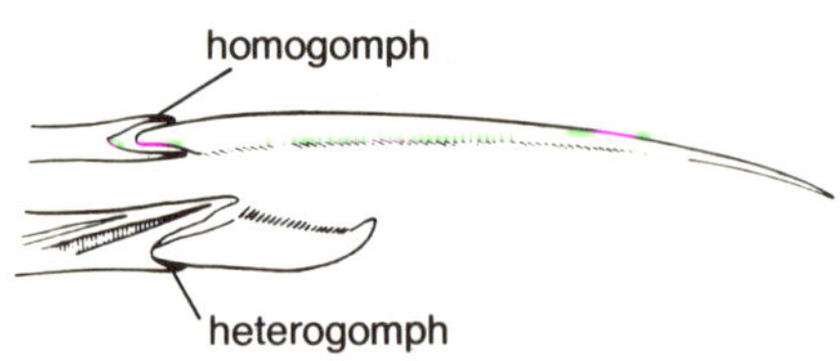

Compound setae

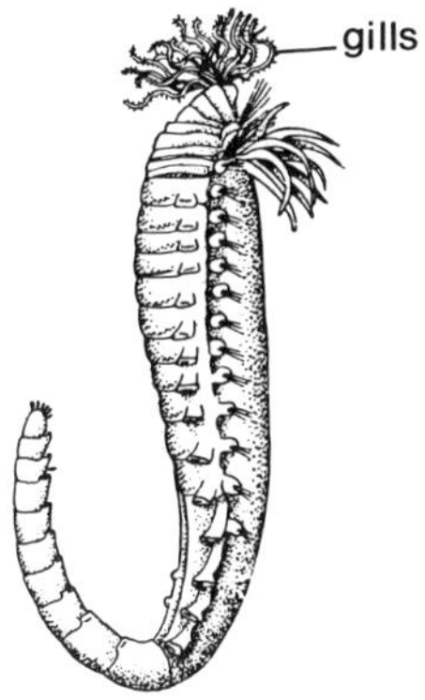

Ampharetidae

Arabellidae

Flabelligeridae

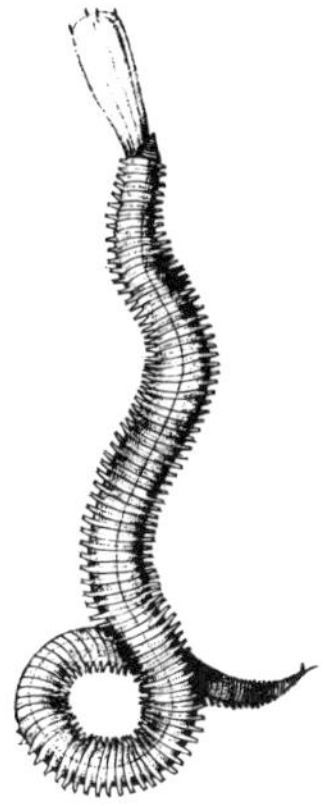
Glyceridae

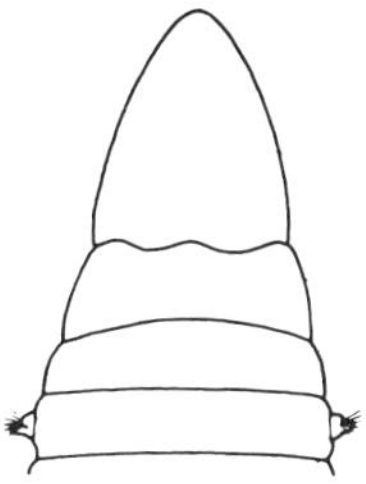
Lumbrineridae

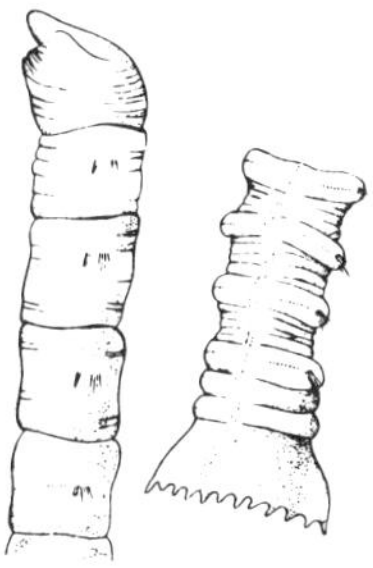
Maldanidae

Nephtyidae

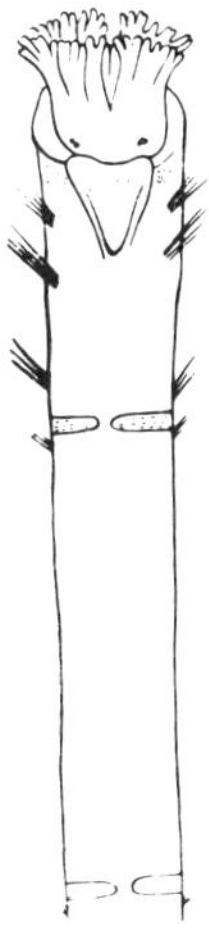
Oweniidae

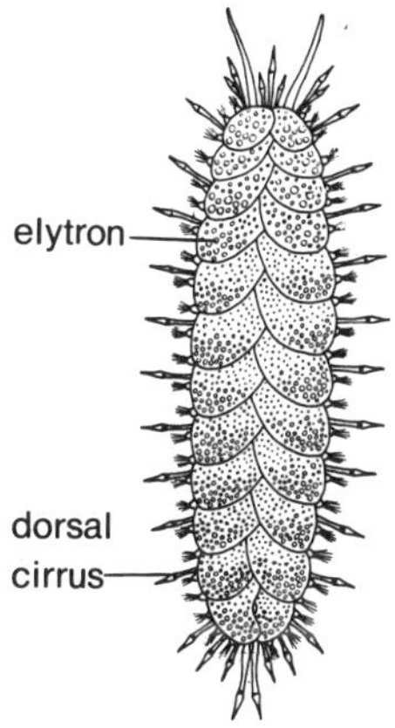

Polynoidae

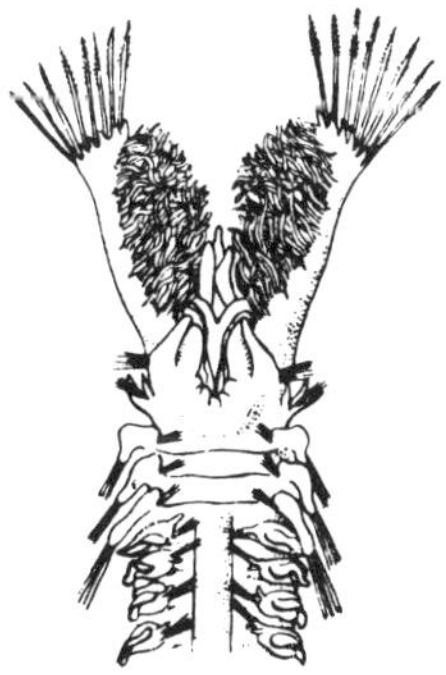
Sabellariidae

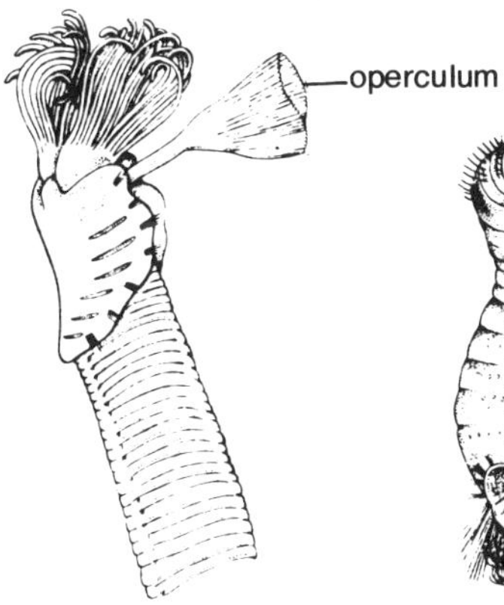

Serpulidae

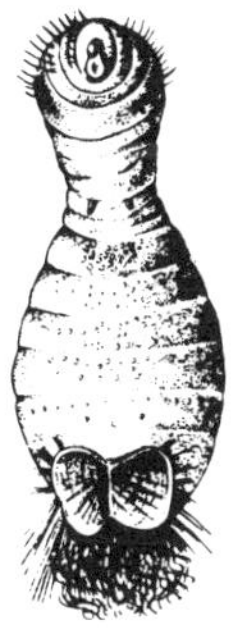
Sternaspidae

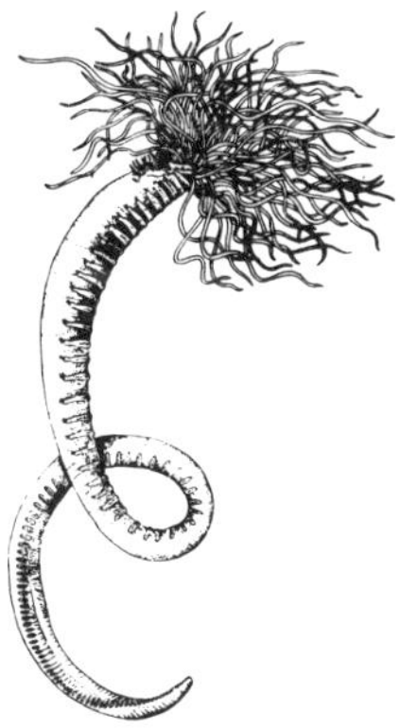
Terebellidae

segments) situated behind the rows of notosetae; prostomial antennae visible only in ventral view (subtidal and rare) Euphrosinidae

20b Notosetae not forming transverse rows that reach nearly to the midline; gills, when present, usually medial to the notosetae, rather than behind them; prostomial antennae visible in dorsal view 21

21a Prostomium continued posteriorly on the dorsal surface as an elongated caruncle (this extends backward for at least 2 or 3 segments and its texture is generally very unlike that of the segments which it interrupts) (subtidal and rare) Amphinomidae

21b Prostomium not continued posteriorly on the dorsal surface as an elongated caruncle 22

22a Head either with long slender setae that form a "cage" around it, or with 1 or more rows of heavy setae that are part of an apparatus for closing the tube in which the animal lives 23

22b Head without setae forming a cage around it or contributing to an apparatus that closes the tube 25

23a Setae at the anterior end long and slender, forming a "cage" around the head Flabelligeridae

23b Setae of the head region short and stout, forming part of an apparatus for closing the tube 24

24a Inhabiting a slender conelike tube, open at both ends and consisting mostly of sand grains cemented together; mobile and living on sandy substrates Pectinariidae

24b Inhabiting coiled or sprawling tubes open only at 1 end and consisting of sand grains and bits of shell embedded in a concrete-like mixture; affixed permanently to rock or shell Sabellariidae

25a With several to many more or less equal tentacles around the mouth (if numerous, these tentacles are long, extensile, and incapable of being retracted into the mouth; if few, they are short and capable of being retracted into the mouth) 26

25b Without several to many more or less equal tentacles around the mouth (however, there may be a pair of long grooved palps, or 2 to several antennae or tentacular cirri arising from the head region, and there may be extensile threadlike tentacles on at least some segments behind the head) 27

26a Tentacles around the mouth long, threadlike, and very extensile, not capable of being withdrawn into the mouth; with 1 or more pairs of simple or branched gills arising on successive segments behind the head; mostly over 5 cm long Terebellidae

26b Tentacles around the mouth short, smooth or somewhat warty, capable of being withdrawn into the mouth; with 2 groups of several simple gills arising just behind the head; mostly under 5 cm long Ampharetidae

27a With a pair of long, grooved palps arising from the head (when these palps are present, they arise from the second segment of the head, but the reduction of the prostomium or fusion of it with the buccal segment may make it appear that the palps come from the prostomium) 28

27b Without a pair of long, grooved palps arising from the head (but there may be short palps or antennae on the prostomium and tentacular cirri arising from the buccal segment behind it) 30

28a Body divided into 3 distinct regions, the segments of the second region being most conspicuously modified Chaetopteridae

28b Body not divided into 3 distinct regions 29

29a With threadlike filaments arising from several segments behind the head Cirratulidae (in part)

29b Without threadlike filaments arising from segments behind the head Spionidae

30a With long threadlike filaments arising from at least several segments behind the head, and sometimes from most segments; without antennae, palps, or tentacular cirri on the head Cirratulidae (in part)

30b Without long threadlike filaments arising from segments behind the head (but there may be short gills); with antennae, palps, or tentacular cirri on the head 31

31a Prostomium conical, tapering to a point and tipped with 4 small tentacles 33

32a Proboscis with four hooklike jaws (to see this character, it may be necessary to induce eversion of the proboscis by applying gentle pressure behind the head) Glyceridae

32b Proboscis with numerous small jaws (there may also be some V-shaped denticles on each side of the proboscis near its base) Goniadidae

33a Prostomium with a pair of palps that are divided into 2 sections (the distal section may be small and nipplelike, and the palp as a whole may be indistinct from the prostomium) 34

33b If palps are present on the prostomium, they are not divided into 2 sections 36

34a Neurosetae all simple Pilargidae

34b Neurosetae compound (however, in some the blade may be secondarily fused to the shaft) 35

35a With 2 antennae (in addition to the palps) on the prostomium: basal sections of palps stout and much larger than the terminal sections; proboscis, when everted, showing 2 ringlike divisions, with a pair of heavy jaws on the anterior division and conical teeth on one or more areas of both rings (to induce eversion of the proboscis, apply gentle pressure just behind the head) Nereidae

35b With 3 antennae (in addition to the palps) on the prostomium; basal sections of palps stout, but smaller than the terminal sections; proboscis not divided into 2 ringlike portions, and without jaws or teeth Hesionidae

36a Prostomium with a pair of palps (but these are sometimes reduced) 37

36b Prostomium without palps 40

37a Prostomium with 3 antennae in addition to the palps; generally with 4 eyes; proboscis with a single tooth or a crown of teeth, or both Syllidae

37b Prostomium with 2, 5, or 7 antennae in addition to the palps; if eyes are present, there are generally 2; proboscis with a system of jaws consisting of 2 smooth lower pieces and 2 or more series of toothed upper pieces 38

38a Only the first segment without parapodia and without setae; with 7 prostomial antennae (the 5 arising near the posterior margin of the prostomium much larger than the 2 arising near the anterior margin); living in tubes Onuphidae

38b Both the first and second segments without parapodia and without setae; with not more than 5 prostomial antennae; living in tubes (Eunicidae) or crawling or burrowing (Dorvilleidae) 39

39a Prostomium with a pair of articulated antennae; palps generally long, curved (in *Ophryotrocha*, which has only 6 setigerous segments, the palps are much reduced); crawling or burrowing Dorvilleidae

39b Prostomium with 5 antennae; palps short, globular, often indistinctly demarcated from the prostomium; living in tubes Eunicidae

40a Prostomium with 4 small antennae, those of the anterior pair being situated at the anterior angles, but without distinct eyes; with a single cirrus at the posterior end; setae never consisting of 2 separate portions; without a group of tentacular cirri on each side of the head behind the prostomium; rather heavy-bodied worms Nephtyidae

40b Prostomium with 4 or 5 tentacles and 2 eyes; with a pair of cirri at the posterior end; setae usually consisting of 2 separate portions; with a group of tentacular cirri on each side of the head just behind the prostomium; generally slender worms Phyllodocidae

Family Ampharetidae

1a With a number of large, anteriorly directed setae arising from the head region 2

1b Without large anteriorly directed setae arising from the head region 5

2a With 14 thoracic setigerous segments *Ampharete arctica* Malmgren, 1865

2b With 15 or 17 thoracic setigerous segments 3

3a With 15 thoracic setigerous segments *Anobothrus gracilis* (Malmgren, 1865)

3b With 17 thoracic setigerous segments 4

4a Gills flattened, those of 1 pair terminating in hooklike tips; large, anteriorly directed setae of head region nearly straight and tapering gradually to fine tips *Amphicteis scaphobranchiata* Moore, 1906

4b Gills terete; large, anteriorly directed setae of head region slightly curved, tapering rather abruptly to fine tips *Amphicteis mucronata* Moore, 1923

5a With 2 types of gills (anterior pair smooth, the other pinnate on 1 or both sides) *Schistocomus hiltoni* Chamberlin, 1919

5b With only 1 type of gill (smooth) *Amage anops* Johnson, 1901

Family Aphroditidae

1a Median antenna at least as long as the prostomium; tips of neurosetae straight or only slightly curved, and with a hood which usually has a frayed or pilose appearance *Aphrodita japonica* Marenzeller, 1879

1b Median antenna decidedly shorter than the prostomium; tips of neurosetae decidedly hooked, and without a hood *Aphrodita negligens* Moore, 1905

Family Arabellidae

1a Parapodia with projecting acicular setae 2

1b Parapodia without projecting acicular setae (prostomium with 4 eyes in a transverse row) *Arabella iricolor* (Montagu, 1804)

2a Posteriormost jaw plates with teeth along the entire length of the inner margin *Notocirrus californiensis* Hartman, 1944

2b Posteriormost jaw plates with teeth only in the basal portion, the rest of each plate being sickle-shaped 3

3a Posterior parapodia with short presetal lobes and elongate postsetal lobes; jaw plates with 2 or 3 teeth in the basal portion *Drilonereis falcata* Moore, 1911

3b Posterior parapodia with elongate presetal and postsetal lobes; posteriormost jaw plates with 3 to 5 teeth in the basal portion

Drilonereis longa Webster, 1879

Family Arenicolidae

1a Relatively large worms, often exceeding a length of 10 cm and a width of 1 cm; not secreting tubes; posterior third of body, behind the region bearing gills, without setae, and therefore taillike; gills pinnate or extensively arborescent 2

1b Small worms, up to about 2.5 cm long and 0.5 mm wide; secreting membranous tubes coated with fine sand; gills and setae continuing nearly to the posterior end of the body, so that there is no taillike portion; gills consisting of 1 to 4 simple filaments
Branchiomaldane vincenti Langerhans, 1881

2a Nephridiopores completely exposed; with 4 to 7 pairs of esophageal caeca; typically inhabiting muddy sand at the inner margins of bays
Abarenicola pacifica Healy & Wells, 1959

2b Ventral half of nephridiopores covered by a flap of skin; with 8 or more pairs of esophageal caeca; typically inhabiting rather clean sand at the mouths of bays or on beaches exposed to some wave action 3

3a Usually with 8 to 10 pairs of esophageal caeca; usually on beaches exposed to some wave action
Abarenicola claparedii oceanica Healy & Wells, 1959

3b Usually with 12 to 19 pairs of esophageal caeca; usually near the mouths of sandy bays
Abarenicola claparedii vagabunda Healy & Wells, 1959
(There may be some overlap in the number of esophageal caeca in the 2 subspecies of *A. claparedii.*)

Family Capitellidae

1a Apparent first segment (it is actually the second, because the peristomium is fused with the prostomium) with setae
Capitella capitata (Fabricius, 1780)

1b Apparent first segment without setae 2

2a Prostomium with ocelli; thoracic region with capillary setae only; without distinct gills in the abdominal region
Notomastus (Clistomastus) tenuis Moore, 1909

2b Prostomium without ocelli; thoracic region with both capillary setae and hooded hooks; with filiform gills beginning on about the fifteenth abdominal segment, the number of gills in each group increasing posteriorly *Heteromastus filobranchus* Berkeley & Berkeley, 1932

Family Chaetopteridae

1a Notopodia of middle region of body with only 1 lobe
Mesochaetopterus taylori Potts, 1914

1b Notopodia of middle region of body bilobed 2

2a With 1 pair of tentacular cirri arising from the dorsal side of the peristomium; tubes transparent or translucent and annulated throughout
Telepsavus costarum Claparède, 1868

2b With 2 pairs of tentacular cirri (those of 1 pair being much smaller than those of the other) arising from the dorsal side of the peristomium; tubes translucent and annulated only in the recently secreted portions *Phyllochaetopterus prolifica* Potts, 1914

Family Chrysopetalidae

Paleanotus bellis (Johnson, 1897) (evidently distinct from *Paleanotus chrysolepis* Schmarda, 1861) is the only species likely to be found.

Family Cirratulidae

1a With a pair of large tentacular cirri arising from the head region 4
1b Without a pair of large tentacular cirri arising from the head region 2
2a With a long unpaired tentacle arising from the dorsal surface in the region of setigerous segments 2 and 3; without parapodial gills; without paired groups of filaments arising on the dorsal surface of the first setigerous segment *Cossura longocirrata* Webster and Benedict, 1887
2b Without a long unpaired tentacle arising from the dorsal surface in the region of setigerous segments 2 and 3; with parapodial gills; with paired groups of filaments arising on the dorsal surface of the first setigerous segment 3
3a With 2 groups of 2 to 8 grooved filaments arising from the first setigerous segment *Cirratulus cirratus cirratus* (O. F. Müller, 1776)
3b With 2 groups of 8 to 20 grooved filaments arising from the first setigerous segment *Cirratulus cirratus spectabilis* (Kinberg, 1866)
4a With 3 to 11 pairs of gills 5
4b With many pairs of gills 6
5a With 6 to 11 pairs of gills; tentacular cirri of the head region often shorter than the first pair of gills *Dodecaceria fewkesi* Berkeley and Berkeley, 1954
5b With 3 to 5 pairs of gills; tentacular cirri of head region as long as or longer than the first pair of gills *Dodecaceria concharum* Oersted, 1843
6a Setae all of the capillary type *Tharyx multifiliis* Moore, 1909
6b Heavy setae present on parapodia of the posterior region *Chaetozone setosa* Malmgren, 1867

Family Dorvilleidae

1a Palps of prostomium poorly developed, resembling small papillae; adults with only 6 setigerous segments *Ophryotrocha vivipara* Banse, 1963
1b Palps of prostomium well developed and usually rather long, a small distal portion sometimes set off from the basal portion 2
2a Parapodia with 2 acicular setae, the upper of these supporting the notopodial lobe, which bears a dorsal cirrus *Dorvillea* spp. See Banse & Hobson (1974) for differentiation of the species.
2b Parapodia with only 1 acicular seta, the notopodial lobe essentially lacking, although a dorsal cirrus is present *Protodorvillea gracilis* (Hartman, 1938)

Family Eunicidae

1a With a pair of tentacular cirri 2
1b Without tentacular cirri *Marphysa stylobranchiata* Moore, 1909
2a Gills beginning on the third setigerous segment, with up to 16 fila-

ments where best developed *Eunice valens* (Chamberlin, 1919)

2b Gills beginning on the fifth to seventh setigerous segment, with from about 15 to 40 filaments where best developed *Eunice aphroditois* (Pallas, 1788)

This is the more common of the 2 species. In recent years, specimens from our region have generally been referred to *E. kobiensis* McIntosh, 1885 (see Gustus, 1972). However, *E. valens* is characterized by having a shorter median antenna, and is probably distinct from *E. kobiensis.*

Family Flabelligeridae

1a Body enclosed by a mucous sheath; neurosetae composite *Flabelligera infundibularis* (Johnson, 1901)

1b Body not enclosed by a mucous sheath; neurosetae simple 2

2a With a pair of prominent nephridial papillae on the ventral side of the fourth or fifth segment; notosetae of anterior segments not forming a cephalic "cage" *Brada villosa* (Rathke, 1843)

2b Without prominent papillae on the ventral side of the fourth or fifth segment; elongated notosetae of anterior segments forming a cephalic "cage" 3

3a Notosetae of cephalic cage long and conspicuous; dermal papillae decidedly elevated *Pherusa papillata* (Johnson, 1901)

3b Notosetae of cephalic cage short and inconspicuous; dermal papillae low *Pherusa plumosa* (O. F. Müller, 1776)

P. papillata and *P. plumosa* are very similar, and it has been suggested that they should be treated as variants of 1 species.

Family Glyceridae

1a All parapodia uniramous *Hemipodus borealis* Johnson, 1901

1b All parapodia, except the first 2, biramous 2

2a Parapodia with only 1 postsetal lip *Glycera capitata* Oersted, 1843

2b Parapodia with 2 postsetal lips 3

3a With compound and retractile gills arising on the posterior side of the bases of the parapodia, beginning at about the sixteenth parapodium; up to about 35 cm long and with 300 or more segments *Glycera americana* Leidy, 1855

3b Without gills arising at the bases of the parapodia; up to about 11 cm long and with about 140 segments *Glycera tesselata* Grube, 1863

Family Goniadidae

1a Proboscis with 2 rows of v-shaped denticles ("chevrons") near its base *Goniada brunnea* Treadwell, 1906

1b Proboscis without v-shaped denticles 2

2a Presetal lobes of parapodia, in region of twenty-fifth parapodium, broadly heart-shaped, with a tonguelike prolongation arising in the cleft *Glycinde armigera* Moore, 1911

2b Presetal lobes of parapodia never heart-shaped, those in the region of the twenty-fifth parapodium merely becoming more narrow distally *Glycinde picta* Berkeley, 1927

Family Hesionidae

1a With 2 prostomial antennae 2
1b With 3 prostomial antennae 3
2a With 6 pairs of tentacular cirri *Micropodarke dubia* (Hessle, 1925)
2b With 8 pairs of tentacular cirri
Kefersteinia cirrata (Keferstein, 1863)
3a With 6 pairs of tentacular cirri (a dark brown species abundant intertidally, especially on muddy sand where there are growths of *Ulva*, *Enteromorpha*, and other algae) *Ophiodromus pugettensis* (Johnson, 1901)
3b With 8 pairs of tentacular cirri
Gyptis brevipalpa (Hartmann-Schröder, 1959)

Family Lumbrineridae

1a With compound (even if imperfectly so) hooked setae in the anterior region (setigerous segments 7 to 30) 2
1b Without compound hooked setae in the anterior region (all hooked setae simple) 4
2a Prostomium rounded *Lumbrineris inflata* Moore, 1911
2b Prostomium pointed 3
3a Both the presetal and postsetal lobes of parapodia of posterior region elongated, several times as long as wide
Lumbrineris cruzensis Hartman, 1944
3b Postsetal lobe of parapodia of posterior region 2 or 3 times as long as wide, the presetal lobe very short and rounded
Lumbrineris latreilli Andouin & Milne-Edwards, 1833
4a Acicula dark brown or black 5
4b Acicula yellow 6
5a Both lobes of parapodia of posterior region of approximately equal length *Lumbrineris bifurcata* (McIntosh, 1885)
5b Postsetal lobe of parapodia of posterior region distinctly longer than the presetal lobe *Lumbrineris similabris* (Treadwell, 1926)
6a Length often exceeding 20 cm; postsetal lobe of parapodia of posterior region only 2 or 3 times as long as wide
Lumbrineris zonata (Johnson, 1901)
6b Length not exceeding 5 cm; postsetal lobe of parapodia of posterior region up to about 10 times as long as wide
Lumbrineris luti Berkeley & Berkeley, 1945

Family Magelonidae

Magelona japonica Okuda, 1937 is the species most likely to be found.

Family Maldanidae

1a Second and third segments with anteriorly directed thin dermal collars, and more posterior segments (beginning with about segment 17) with posteriorly directed collars *Rhodine bitorquata* Moore, 1923
1b None of the segments with thin dermal collars 2
2a Prostomium bearing a plate with a leathery rim 3
2b Prostomium not bearing a plate with a leathery rim (the prostomium may, however, have a distinctly flattened area)
Nicomache lumbricalis (Fabricius, 1780)

3a Posterior end of body terminating in a nearly flat plate, the anus being dorsal in position *Maldane glebifex* Grube, 1860

3b Posterior end of body terminating in a cone surrounded by cirri, the anus being situated on the cone 4

4a First 3 setigerous segments with acicular setae in place of hooklike setae characteristic of more posterior segments 5

4b First 3 setigerous segments with hooklike setae of the type characteristic of more posterior segments *Axiothella rubrocincta* (Johnson, 1901)

5a Base of anal cone situated rather deep within the funnellike region from which the posterior cirri extend *Euclymene zonalis* Verrill, 1874

5b Base of anal cone situated close to the rim of the funnellike region from which the posterior cirri extend *Praxillella gracilis* (Sars, 1861)

Family Nephtyidae

1a Second pair of prostomial antennae bifurcated *Nephtys cornuta* Berkeley & Berkeley, 1945

1b Second pair of prostomial antennae simple 2

2a Gills between the notopodia and neuropodia beginning on tenth or eleventh setigerous segment *Nephtys punctata* Hartman, 1938

2b Gills between the notopodia and neuropodia beginning on the third to sixth setigerous segment 3

3a Notopodia at least somewhat bilobed, the cleft being at the point where the aciculum emerges 4

3b Notopodia rather uniformly rounded, without a cleft where the aciculum emerges *Nephtys longosetosa* Oersted, 1843

4a Proximal portion of proboscis rough 5

4b Proximal portion of proboscis smooth *Nephtys caecoides* Hartman, 1938

5a Proboscis with a fingerlike dorsal papilla near the distal end *Nephtys ciliata* (O. F. Müller, 1776)

5b Proboscis without a median dorsal papilla *Nephtys caeca* (Fabricius, 1780)

Family Nereidae

1a Peristomium enlarged, forming a collar around and under the prostomium; usually commensal with hermit crabs *Cheilonereis cyclurus* (Harrington, 1897)

1b Peristomium not forming a collar around and under the prostomium; not regularly associated with hermit crabs 2

2a Proboscis with paragnaths (small denticles) in addition to the pair of large jaws 3

2b Proboscis without paragnaths *Micronereis nanaimoensis* Berkeley & Berkeley, 1953

3a Paragnaths present in all areas (I-VIII) of the proboscis; length attaining 80 or 90 cm *Nereis (Neanthes) brandti* (Malmgren, 1866)

3b Paragnaths absent in at least area I (mid-dorsal on maxillary distal ring) and/or area V (mid-dorsal on oral [proximal] ring); length not exceeding 25 cm 4

4a Paragnaths absent on oral (proximal) ring of proboscis *Nereis (Ceratonereis) paucidentata* Moore, 1903

4b Paragnaths present on oral ring of proboscis 5

5a Paragnaths on the whole distinct, well developed, and separate from one another 6
5b Paragnaths small and pale, most of them arranged in pectinate rows in which the individual paragnaths are not very distinct *Platynereis bicanaliculata* (Baird, 1863)
6a Notopodia of posterior region of body without homogomph falcigerous setae *Nereis (Neanthes) limnicola* Johnson, 1903
6b Notopodia of posterior region with homogomph falcigerous setae 7
7a Peristomium elongated, about two and a half times as long as wide *Nereis (Nereis) grubei* (Kinberg, 1866)
7b Peristomium not elongated, wider than long 8
8a Upper lobe of notopodia of posterior region considerably enlarged, straplike, several times as long as the lower lobe *Nereis vexillosa* Grube, 1851
8b Upper lobe of notopodia of posterior region not particularly enlarged, not more than about twice as long as the lower lobe 9
9a Upper lobe of notopodia of posterior region nearly rectangular and much larger than the lower lobe *Nereis (Nereis) mediator* Chamberlin, 1918
9b Upper lobe of notopodia of posterior region neither nearly rectangular nor markedly larger than the lower lobe 10
10a Two pairs of tentacular cirri of peristomium decidedly shorter than the other 2 pairs *Nereis (Nereis) procera* Ehlers, 1869
10b All 4 pairs of tentacular cirri of peristomium of about equal length *Nereis (Nereis) pelagica* Linnaeus, 1767

Family Onuphidae

1a Gills simple 2
1b Gills with many branches, these arranged spirally on the main stem *Diopatra ornata* Moore, 1911
2a Gills beginning on the first setigerous segment 3
2b Gills beginning on the third to sixth setigerous segment *Onuphis geophiliformis* (Moore, 1903)
3a Gills and dorsal cirri of typical parapodia (segments 20 to 30) slender throughout, tapering gradually *Onuphis iridescens* (Johnson, 1901)
3b Gills of typical parapodia (segments 20 to 30) thick at the base and tapering rather abruptly, the basal portion of the dorsal cirri similarly thick, supporting a more slender terminal portion of about the same length *Onuphis elegans* (Johnson, 1901)

Family Opheliidae

1a Without a ventral groove; stout, up to about 85 mm long and 20 mm or more wide; with a very characteristic odor reminiscent of garlic or an unlighted gas jet *Travisia pupa* Moore, 1906
1b With a distinct ventral groove; relatively slender worms, the width not exceeding about 5 mm; without any characteristic odor 2
2a With 29 setigerous segments; with lateral eyes between the parapodia from setigerous segment 7 to 17 *Armandia brevis* (Moore, 1906)
2b With 42 to 50 setigerous segments; without lateral eyes between the parapodia *Ammotrypane aulogaster* Rathke, 1843

Family Orbiniidae

1a Prostomium rounded or truncate, not distinctly pointed 2
1b Prostomium pointed 3
2a Prostomium rounded; proboscis, when everted, with distinctly branched lobes *Nainereis dendritica* (Kinberg, 1867)
2b Prostomium broadly truncate; proboscis, when everted, somewhat lobed, but the lobes do not branch *Nainereis berkeleyorum* Pettibone, 1957
3a Thoracic neuropodia with blunt, toothed setae in addition to capillary setae *Scoloplos armiger* (O. F. Müller, 1776)
3b Thoracic neuropodia with only capillary setae *Haploscoloplos elongatus* (Johnson, 1901)

Family Oweniidae

1a Prostomium with several branched lobes *Owenia fusiformis* delle Chiaje, 1844
1b Prostomium without branched lobes *Myriochele heeri* Malmgren, 1867

Family Pectinariidae

1a Conspicuous teeth of hooklike setae arranged in a single row *Pectinaria (Cistenides) granulata* (Linnaeus, 1767)
1b Conspicuous teeth of hooklike setae arranged in a double row *Pectinaria (Pectinaria) californiensis* Hartman, 1941

Family Phyllodocidae

1a With 2 pairs of tentacular cirri, these arising from 1 segment 2
1b With 3 or 4 pairs of tentacular cirri, these arising from 2 or 3 segments 3
2a Dorsal cirri of parapodia thick, conical in outline, nearly symmetrical, about as long as wide *Eteone longa* (Fabricius, 1780)
2b Dorsal cirri of parapodia thin, rounded distally, asymmetrical, much wider than long *Eteone pacifica* Hartman, 1936
For other species of *Eteone* that may be encountered, see Banse & Hobson (1974).
3a With 3 pairs of tentacular cirri arising from 2 segments (the more nearly ventral pair on segment 2 may be short) 4
3b With 4 pairs of tentacular cirri arising from 3 (or apparently 2) segments (the more nearly ventral pair on segment 2 may be leaflike) 5
4a Segment 2 with setae and an elongate ventral cirrus *Mystides borealis* Theel, 1879
4b Segment 2 without setae (setae begin on segment 3) and with a short, broad ventral cirrus *Hesionura coineaui difficilis* (Banse, 1963)
5a Prostomium with a median antenna 6
5b Prostomium without a median antenna 7
6a Parapodia essentially with 2 lobes; notosetae simple, neurosetae compound 9
6b Parapodia with only 1 lobe; all setae compound *Eulalia* spp.
Eulalia (Bergstroemia) nigrimaculata Moore, 1909 is perhaps our most common intertidal species. For other species, belonging to various subgenera, see Banse & Hobson (1974).
7a Ventral cirri with an acuminate ventral projection

Phyllodoce (Anaitides) groenlandica Oersted, 1843

7b Ventral cirri without an acuminate ventral projection 8

8a Ventral cirri oval; with a brown spot on each dorsal cirrus and in the median region of each segment, these spots collectively forming 3 longitudinal bands *Phyllodoce (Anaitides) maculata* (Linnaeus, 1767)

8b Ventral cirri lanceolate; with longitudinal white bands on the dorsal surface in the posterior part of the body *Phyllodoce (Anaitides) mucosa* Oersted, 1843

For other species of *Phyllodoce* that may be encountered, see Banse & Hobson (1974).

9a The more nearly ventral tentacular cirri on segment 2 slender, not leaflike *Notophyllum (Notophyllum) imbricatum* Moore, 1906

9b The more nearly ventral tentacular cirri on segment 2 leaflike *Notophyllum (Hesperophyllum) tectum* Chamberlin, 1919

Family Pilargidae

1a Acicular setae bent distally and protruding from the notopodia of the middle region of the body *Sigambra tentaculata* (Treadwell, 1941)

1b Acicular setae not usually bent distally and not protruding from the notopodia 2

2a Median prostomial antenna absent; body surface papillated *Pilargis berkeleyae* Monro, 1933

2b Median prostomial antenna present; body surface smooth or wrinkled, but not papillated *Otopsis longipes* Ditlevsen, 1917

Family Polynoidae

1a With 12 pairs of elytra *Lepidonotus squamatus* (Linnaeus, 1767)

1b With at least 15 pairs of elytra 2

2a With 15 pairs of elytra 4

2b With at least 18 pairs of elytra 3

3a With 18 pairs of elytra *Halosydna brevisetosa* Kinberg, 1855

3b With more than 18 pairs of elytra 18

4a Body with more than 50 segments; elytra restricted to the anterior region, at least 18 of the posterior segments being uncovered 5

4b Body with fewer than 50 segments; elytra almost completely covering the body 7

5a Notosetae stouter than the neurosetae *Hermadion truncata* (Moore, 1902)

5b Notosetae more slender than the neurosetae 6

6a Prostomium with prominent cephalic peaks; with 2 mid-dorsal knobs on each segment; elytra colorless *Enipo canadensis* (McIntosh, 1874)

6b Prostomium rounded anteriorly, without cephalic peaks; segments without mid-dorsal knobs; elytra with a smoky brown patch *Polynoe gracilis* (Verrill, 1874)

7a At least some of the neurosetae bifid at the tip 8

7b None of the neurosetae bifid at the tip 12

8a Elytra smooth except for a group of small tubercles at the anterior curved part; notosetae with only faint rows of spines *Harmothoe lunulata* (delle Chiaje, 1841)

8b Elytra with small tubercles covering most of the surface; notosetae with distinct rows of spines 9

9a Anterior pair of eyes (on dorsolateral surfaces of prostomium) visible from the dorsal side 11

9b Anterior pair of eyes (on anteroventral surfaces of prostomium) not visible from the dorsal side *Harmothoe imbricata* (Linnaeus, 1766)

10a Soft macrotubercles of elytra wide at the base, not sharply demarcated from the elytral surface; elytra with slightly curved sharp-pointed tubercles intermediate in size between the macrotubercles and microtubercles *Lagisca multisetosa* Moore, 1902

10b Soft macrotubercles of the elytra narrow at the base, thus sharply demarcated from the elytral surface; elytra without sharp-pointed tubercles intermediate in size between macrotubercles and microtubercles *Harmothoe extenuata* (Grube, 1840)

11a Some notosetae with capillary tips 13

11b None of the notosetae with capillary tips 12

12a Elytra cream-colored; microtubercles of elytra conical; slender papillae not regularly present on elytral margins; papillae on antennae and cirri short *Eunoe depressa* Moore, 1905

12b Elytra mottled with brown or gray; at least some microtubercles of elytra pronged; slender papillae regularly present on elytral margins; papillae on antennae and cirri long *Eunoe oerstedi* Malmgren, 1865

13a Neurosetae of 2 types, those above the acicula more slender than those below the acicula *Hesperonoe complanata* (Johnson, 1901)

13b Neurosetae of only 1 type 14

14a Lateral prostomial antennae inserted ventrally; at least some of the microtubercles of the elytra pronged 15

14b Lateral prostomial antennae inserted subterminally; microtubercles of elytra not pronged *Gattyana iphionelloides* (Johnson, 1901)

15a All notosetae with fine capillary tips; elytra with large conical macrotubercles, these with truncate, roughened summits; cephalic peaks of prostomium short and blunt *Gattyana ciliata* Moore, 1902

15b Notosetae of 2 types, the upper ones stout and blunt, the lower ones slender and with capillary tips; elytra without large conical macrotubercles; prostomium with distinctly projecting cephalic peaks 16

16a Microtubercles of elytra with as many as 4 prongs; bare tips of lower neurosetae not longer than the spinous regions *Gattyana cirrosa* (Pallas, 1766)

16b Some microtubercles of elytra with 8 or more prongs; bare tips of lower neurosetae longer than the spinous regions *Gattyana treadwelli* Pettibone, 1949

17a Elytra on every segment posterior to segment 38; dorsal nuchal fold prominent, and covering part of the prostomium *Hololepida magna* Moore, 1905

17b Elytra not on every segment posterior to segment 38; without a prominent nuchal fold 18

18a Lateral prostomial antenna inserted terminally on the prostomium; notopodia abortive, with an aciculum but without notosetae 19

18b Lateral prostomial antenna inserted terminoventrally; notosetae present, although not necessarily on all notopodia, and sometimes few in number 20

19a Elytra of each pair large enough to touch or nearly touch medially; middle neurosetae with the clearly bifid and smooth terminal portion extending some distance beyond the portion bearing spines *Lepidasthenia longicirrata* Berkeley, 1923

19b Elytra of each pair so small that they are widely separated; middle neurosetae with the smooth terminal portion not extending beyond the portion bearing spines, and if the tip is bifid, the smaller tooth may

appear essentially as 1 of the spines
Lepidasthenia berkeleyae Pettibone, 1948

20a Elytral pairs on segments 2, 4, 5, 7, 9 . . . 23, 26, 29, 32, 35; following segment 35, the arrangement of elytra is irregular because they are not in sequence and not always in pairs; notosetae numerous, with pointed tips *Polyeunoa tuta* (Grube, 1855)

20b Elytral pairs on segments 2, 4, 5, 7, 9 . . . 23, 26, 29, 31, 33, then every other segment to the end of the body; relatively few segments near the posterior end with unpaired elytra; notosetae few with rather blunt, indistinctly bifid tips, or absent 21

21a Margins of elytra extensively ruffled or folded
Arctonoe fragilis (Baird, 1863)

21b Margins of elytra smooth or only slightly wavy 22

22a Neurosetae (except on segment 2) of 2 types, those above the aciculum slender and with blunt, bifid tips, those below the aciculum stouter and with falcate, pointed tips; with a dark band of pigment across segments 7 to 8 *Arctonoe vittata* (Grube, 1855)

22b Neurosetae all essentially of 1 type, falcate and with pointed tips; without a dark band of pigment across segments 7 to 8
Arctonoe pulchra (Johnson, 1897)

Family Polyodontidae

Peisidice aspera Johnson, 1897 is the only species likely to be found in our area.

Family Sabellariidae

1a Opercular paleae arranged in 2 rows *Idanthyrsus armatus* Kinberg, 1867

1b Opercular paleae arranged in 3 rows *Sabellaria cementarium* Moore, 1906

Family Sabellidae

1a Prostomial cirri dichotomously branched 1 or more times
Schizobranchia insignis Bush, 1904

1b Prostomial cirri not dichotomously branched 2

2a Two to 4 prostomial cirri on either side with conspicuous swollen ocelli curled spirally about them near their tips
Megalomma splendida (Moore, 1905)

2b Prostomial cirri without conspicuous ocelli near their tips (however, smaller ocelli may be scattered over the cirri) 3

3a Hooklike setae of thoracic region with stems several times the length of the terminal dentate portion 7

3b Hooklike setae of thoracic region with short stems or with no apparent stems 4

4a Prostomial lobes coiled in a spiral
Eudistylia vancouveri (Kinberg, 1867)

4b Prostomial lobes not coiled in a spiral 5

5a Neuropodia of thoracic region with only hooklike setae
Laonome krøyeri Malmgren, 1865

5b Neuropodia of thoracic region with both hooklike setae and setae with picklike terminal blades 6

6a Notopodia of thoracic region provided only with setae having elongated blades *Sabella crassicornis* M. Sars, 1851

6b Notopodia of thoracic region provided both with setae having elongated blades and those with spatulate or nearly spatulate blades
Pseudopotamilla occelata (Moore, 1905)

7a Secreting thick, nearly transparent, slimy tubes; hooklike notosetae of abdominal region nearly encircling the latter
Myxicola infundibulum (Renier, 1804)

7b Secreting opaque, leathery tubes; hooklike setae of abdominal region forming only short series 8

8a Prostomial cirri entirely separate to their bases
Fabricia sabella oregonica Banse, 1956

8b Prostomial cirri partially united by a membrane 9

9a With a wide ventral anal depression; with about 10 prostomial cirri arising from each lobe *Euchone rosea* Langerhans, 1884

9b Without a ventral anal depression; with 12 to 36 prostomial cirri arising from each lobe *Chone infundibuliformis* Krøyer, 1856

Family Scalibregmidae

Scalibregma inflatum Rathke, 1843 is the only species likely to be encountered in our area.

Family Serpulidae

1a Body symmetrical, with more than 5 setigerous segments in the thoracic region 2

1b Body asymmetrical, with fewer than 5 setigerous segments in the thoracic region 4

2a Operculum funnel-shaped 3

2b Operculum nearly globular *Chitinopoma groenlandica* (Mörch, 1863)

3a Basal portion of operculum with 2 rounded basal processes, which in turn may be bilobed *Crucigera irregularis* Bush, 1904

3b Basal portion of operculum without rounded basal processes
Serpula vermicularis Linnaeus, 1767

4a Tube coiled clockwise from the aperture to the apex 5

4b Tube coiled counterclockwise from the aperture to the apex 6

5a With 3 setigerous segments in the thoracic region
Spirorbis (Dexiospira) spirillum (Linnaeus, 1758)

5b With 4 setigerous segments in the thoracic region (the fourth is incomplete, having only hooked setae, and no setae at all on the notopodium) *Spirorbis (Paradexiospira) vitreus* (Fabricius, 1780)

6a Brooding in the tube 7

6b Brooding in the operculum 8

7a With 4 setigerous segments in the thoracic region; collar setae with a crenulated basal expansion
Spirorbis (Protolaeospira) amphilateralis Pixell, 1912

7b With 3 setigerous segments in the thoracic region; collar setae without a crenulated basal expansion
Spirorbis (Laeospira) medius Pixell, 1912

8a Opercular plate bilobed; blades of collar setae coarsely serrated
Spirorbis (Laeospira) mörchi (Levinsen, 1884)

8b Opercular plate not bilobed; blades of collar setae finely serrated
Spirorbis (Laeospira) granulatus (Linnaeus, 1767)

Family Sigalionidae

1a Each segment from segment 27 posteriorly bearing elytra; basal portion of median prostomial antenna with a pair of lateral palplike outgrowths; ciliated cirriform gills arising from lateral borders of elytrophores and dorsal tubercles 2

1b Each segment from segment 23 posteriorly bearing elytra; median prostomial antenna without lateral palplike outgrowths; no gills arising from lateral borders of elytrophores or dorsal tubercles *Pholoe minuta* (Fabricius, 1780)

2a Body, exclusive of setae, up to 7 to 10 mm wide; ventral surface with short close-set papillae between bases of parapodia; compound neurosetae regularly falcate and with bifid tips *Sthenelais berkeleyi* Pettibone, 1971

2b Body, exclusive of setae, up to 4 or 5 mm wide; ventral surface without papillae; compound neurosetae mostly straight and tapering to slender tips, only a few of them falcate and with bifid tips *Sthenelais tertiaglabra* Moore, 1910

Family Sphaerodoridae

1a Body with fewer than 30 segments; dorsal surface with 8 to 14 longitudinal rows of globular papillae; all setae compound 2

1b Body with more than 50 segments; dorsal surface with 2 longitudinal rows of globular papillae; all setae simple *Sphaerodorum papillifer* Moore, 1909

2a Dorsal surface with numerous small papillae in addition to the large, globular papillae *Sphaerodoropsis minuta* (Webster & Benedict, 1887)

2b Dorsal surface without small papillae *Sphaerodoropsis sphaerulifer* (Moore, 1909)

Family Spionidae

1a Fifth setigerous segment at least somewhat enlarged and provided with specialized heavy setae 2

1b Fifth setigerous segment not noticeably enlarged, without specialized heavy setae 6

2a Gills first present anterior to the fifth setigerous segment *Boccardia* spp.

2b Gills first present posterior to the fifth setigerous segment 3

3a Fifth setigerous segment only slightly modified, and with a prominent post-setal lobe; heavy setae of fifth setigerous segments arranged in a J-shaped double row *Pseudopolydora kempi japonica* Imajima & Hartman, 1964

3b Fifth setigerous segment greatly modified, without a post-setal lobe; heavy setae of fifth setigerous segment usually arranged in a crescent-shaped row 4

4a Gills beginning on the second setigerous segment *Polydora californica* Treadwell, 1914

4b Gills beginning on the sixth or seventh setigerous segment 5

5a Gills beginning on the sixth setigerous segment; commensal within shells occupied by hermit crabs *Polydora commensalis* Andrews, 1891

5b Gills beginning on the seventh setigerous segment; not commensal with hermit crabs *Polydora ligni* Webster, 1879

6a Gills absent *Spiophanes cirrata* Sars, 1892
6b Gills present 7
7a Successive neuropodia in the elongated posterior region connected by inter-ramal pouches *Laonice cirrata* (Sars, 1861)
7b Successive neuropodia of the posterior region of the body not connected by inter-ramal pouches 8
8a Gills found almost throughout the length of the body *Nerine foliosa* (Andouin & Milne-Edwards, 1833)
8b Gills limited to the anterior half of the body 9
9a With 6 to 13 pairs of simple gills *Prionospio cirrifera* Wiren, 1883
9b With 3 or 4 pairs of pinnate gills *Prionospio pinnata* Ehlers, 1901

Family Sternaspidae

Sternaspis fossor Stimpson, 1854 is common subtidally in muddy bottoms.

Family Syllidae

1a Ventral cirri absent; dorsal cirri smooth *Autolytus* spp. Banse and Hobson (1974) have provided a key to asexual forms of several species known to occur in our area. Some of the sexual forms that have been described and given names have not been linked to the asexual forms from which they arise. Thus the taxonomy of this genus is confused.
1b Ventral cirri present; dorsal cirri either smooth or annulated 2
2a Prostomial palps either decidedly separate or united only at their bases; length generally exceeding 1 cm 3
2b Prostomial palps joined completely or almost completely; length generally less than 5 mm 15
3a Prostomial antennae, tentacular cirri, and dorsal cirri beaded or articulated; palps separate or fused at their bases 4
3b Prostomial antennae, tentacular cirri, and dorsal cirri smooth or slightly annulated (usually only distally) (except in *Syllides*, in which the dorsal cirri are beaded); palps fused at their bases 6
4a Body dorsoventrally flattened; pharynx with a number of teeth at its margin, but without a large mid-dorsal tooth 5
4b Body not appreciably flattened; pharynx without a number of teeth at its margin, but with a large mid-dorsal tooth *Syllis* spp. A number of species are found in our area; see Banse & Hobson (1974).
5a Compound setae with a single tooth (except in the stolons involved in asexual budding) *Trypanosyllis (Trypanadenta) ingens* Johnson, 1902
5b Compound setae regularly with 2 teeth *Trypanosyllis (Trypanadenta) gemmipara* Johnson, 1902
6a Length of dorsal cirri less than 5 times the width of the body 7
6b Length of dorsal cirri at least 5 times the width of the body *Amblyosyllis* sp.
7a Pharynx with several large teeth (these are usually directed posteriorly) 8
7b Pharynx with just 1 large mid-dorsal tooth, or with a large tooth plus many small teeth along the margin, or without any teeth at all 10
8a Posterior setigerous segments (after about setiger 40) divided superficially into 2 rings *Odontosyllis fulgurans japonica* Imajima, 1966
8b Posterior setigerous segments not divided into 2 rings 9

9a Compound setae of the anterior parapodia with a fringe of hairs on the concave margin *Odontosyllis phosphorea* Moore, 1909
9b Compound setae of the anterior parapodia without a fringe of hairs on the concave margin *Odontosyllis parva* Berkeley, 1923
10a Pharynx with a mid-dorsal tooth 11
10b Pharynx without a mid-dorsal tooth 13
11a Anterior margin of pharynx finely toothed *Eusyllis* spp.
E. magnifica (Moore, 1906) is probably the species most likely to be encountered; see Banse & Hobson (1974) for a key to the 4 species known to occur in this area.
11b Anterior margin of pharynx not toothed 12
12a All parapodia with compound setae of 1 type only, these with blades more than 3 times as long as wide *Pionosyllis gigantea* Moore, 1908
12b Anterior parapodia with the upper compound setae having very long blades, the lower setae having short blades that are broadened distally *Pionosyllis uraga* Imajima, 1966
13a Acicula of the parapodia of some of the first few segments with knoblike tips; dorsal cirri smooth or irregularly ringed, but not strongly beaded *Streptosyllis latipalpa* Banse, 1968
13b None of the acicula with knoblike tips; dorsal cirri, beginning with setiger 3, strongly beaded 14
14a Upper simple setae of the first few parapodia thick and bent at the tips, differing from the long and slender setae of the remaining parapodia *Syllides longocirrata* Oersted, 1845
14b Upper simple setae of the first few parapodia similar to those of the remaining parapodia *Syllides japonica* Imajima, 1966
15a With 2 pairs of tentacular cirri *Brania brevipharyngea* Banse, 1962
15b With 1 pair of tentacular cirri 16
16a Tentacular cirri well developed; body surface papillated *Sphaerosyllis* spp.
See Banse & Hobson (1974) for a key to the 3 species known to occur in this region.
16b Tentacular cirri small; body surface smooth 17
17a Median prostomial antenna of about the same length as the lateral antennae *Exogone verrugera* (Claparède, 1868)
17b Median prostomial antenna decidedly longer than the lateral antennae 18
18a Median prostomial antenna reaching a very short distance beyond the ends of the palps; tentacular cirri reduced to small knobs; with up to about 33 setigerous segments *Exogone gemmifera* (Pagenstecher, 1862)
18b Median prostomial antenna not reaching to the ends of the palps; tentacular cirri of peristomium ovoid; with up to about 45 setigerous segments *Exogone lourei* Berkeley & Berkeley, 1938
E. molesta Banse, 1972 may also be encountered; see Banse & Hobson (1974).

Family Terebellidae

1a Peristomium forming a large, extensible, papillated proboscis *Artacama conifera* (Moore, 1905)
1b Peristomium not forming an extensible proboscis 2
2a Hooklike setae of the thoracic region different from those of the abdominal region 3
2b Hooklike setae of the thoracic region similar to those of the abdomi-

nal region (exception: in *Pista*, the hooklike setae of the first few thoracic segments are different) 4

3a With a single median gill with 4 pectinate branches *Terebellides stroemi* Sars, 1835

3b With 3 pairs of simple gills *Trichobranchus glacialis* Malmgren, 1865

4a Without gills 5

4b With 1 to 3 pairs of gills 6

5a With 30 to 60 setigerous segments in the thoracic region, beginning with the second segment *Polycirrus caliendrum* Claparède, 1868

5b With 11 to 15 setigerous segments in the thoracic region, beginning with the third segment *Polycirrus kerguelenensis* McIntosh, 1885

6a Hooklike setae of anterior setigerous segments with a posterior prolongation 7

6b Hooklike setae of anterior setigerous segments without a posterior prolongation 10

7a With 1 or 2 pairs (generally 2 pairs) of gills; gills rather compact, the short branches collectively forming a nearly globular mass; length up to about 9 cm *Pista cristata* (O. F. Müller, 1776)

7b With 2 pairs of gills; branches of gills not restricted to a nearly globular mass; length often exceeding 15 cm 8

8a Prostomium with prominent lateral lobes *Pista fasciata* (Grube, 1870)

8b Prostomium without lateral lobes 9

9a Fourth segment with prominent tongue-shaped lateral lobes *Pista pacifica* Berkeley, 1942

9b Fourth segment without tongue-shaped lateral lobes *Pista elongata* Moore, 1909

10a Capillary setae beginning on second segment (first gill-bearing segment) *Streblosoma bairdi* (Malmgren, 1865)

10b Capillary setae beginning on third or fourth segment (second or third gill-bearing segment) 11

11a Capillary setae beginning on third segment (second gill-bearing segment) 12

11b Capillary setae beginning on fourth segment (third gill-bearing segment) except in *Scionella*, which has only 1 pair of gills 14

12a With 2 pairs of gills *Thelepus cincinnatus* (Fabricius, 1780)

12b With 3 pairs of gills 13

13a Hooklike setae always in single rows *Thelepus setosus* (Quatrefages, 1865)

13b Hooklike setae forming elliptical rings, except in the first and last few segments possessing setae of this type *Thelepus crispus* Johnson, 1901

14a Tips of notosetae appearing frayed or ciliated at low magnification 15

14b Tips of notosetae appearing smooth at low magnification 16

15a Gills with a well developed main stem and repeatedly subdivided branches *Neoamphitrite edwardsii* (Quatrefages, 1865)

15b Gills with a very short main stem and undivided branches *Amphitrite cirrata* (O. F. Müller, 1776)

16a With 1 pair of gills; gills dichotomously branched in 1 plane; length up to about 2.5 cm *Scionella japonica* Moore, 1903

16b With 3 pairs of gills; gills arborescent; length up to 12 cm *Eupolymnia heterobranchia* (Johnson, 1901)

Family Tomopteridae

1a With a pair of slender tentacular cirri between the prostomial antennae and the very long cirriform appendages ("streamers") of the head region *Tomopteris renata* E. Berkeley, 1931
T. renata is perhaps synonymous with *T. pacifica* Izuka, 1914.

1b Without a pair of slender tentacular cirri between the prostomial antennae and the very long cirriform appendages ("streamers") of the head region *Tomopteris septentrionalis* Steenstrup, 1849

Banse, K. 1972. On some species of Phyllodicidae, Syllidae, Nephtyidae, Goniadidae, Apistobranchidae, and Spionidae (Polychaeta) from the northeast Pacific Ocean. Pacific Sci., 26:191-222.

Banse, K. & K. D. Hobson. 1968. Benthic polychaetes from Puget Sound, Washington, with remarks on four other species. Proc. U. S. Nat. Mus., 125:1-53.

------. 1974. Benthic errantiate polychaetes of British Columbia and Washington. Bull. 185, Fish. Res. Bd. Canada. 111 pp.

Banse, K. & F. H. Nichols. 1968. Two new species and 3 new records of benthic polychaetes from Puget Sound (Washington). Proc. Biol. Soc. Wash., 81:223-30.

Berkeley, E. 1966. Records of some species of Polychaeta new to British Columbia and of extensions in distribution of some others. Canad. J. Zool., 44:839-49.

------. 1967. A checklist of Polychaeta recorded from British Columbia since 1923, with references to name changes, descriptions, and synonymies. I. Errantia. Canad. J. Zool., 45:1049-59.

------. 1972. Further records of Polychaeta new to British Columbia with comments on some others. Canad. J. Zool., 50:451-6.

Berkeley, E., & C. Berkeley. 1948. Annelida. Polychaeta Errantia. *In* Canadian Pacific Fauna, 9b (1). 100 pp.

------. 1952. Annelida. Polychaeta Sedentaria. *In* Canadian Pacific Fauna, 9b (2). 139 pp.

------. 1957. On some pelagic Polychaeta from the northeast Pacific north of latitude 40° N. and east of longitude 175° W. Canad. J. Zool., 35: 573-8.

------. 1960. Some further records of pelagic Polychaeta from the northeast Pacific north of latitude 40° N. and east of longitude 175° W., together with records of Siphonophora, Mollusca, and Tunicata from the same region. Canad. J. Zool., 38:787-99.

Dales, R. P. 1957. Pelagic polychaetes of the Pacific Ocean. Bull. Scripps Inst. Oceanogr., 7:99-167.

Day, J. H. 1967. A monograph of the Polychaeta of Southern Africa. Parts I (Errantia) and 2 (Sedentaria). London: British Museum (Natural History). xxix + 258 pp.; xvii + 878 pp.

Fauvel, P. 1923. Polychètes Errantes. *In* Faune de France, 5. Paris: Lechevalier. 488 pp.

------. 1927. Polychètes Sedentaires. Addenda aux Errantes, Archiannelides, Myzostomaires. *In* Faune de France, 16. Paris: Lechevalier. 494 pp.

Gustus, R. 1972. A species of the genus *Eunice* (Polychaeta) from the Pacific Northwest coast. Northwest Sci., 46:257-69.

Hartman, O. 1951. Literature of the Polychaetous Annelids. Los Angeles (privately published). 290 pp.

------. 1968. Atlas of the Errantiate Polychaetous Annelids from California. Los Angeles: Allan Hancock Foundation, University of Southern California. 828 pp.

------. 1969. Atlas of the Sedentariate Polychaetous Annelids from California. Los Angeles: Allan Hancock Foundation, University of Southern California. 812 pp.

Hartman, O., & D. J. Reish. 1950. The marine annelids of Oregon. Oregon State Monogr., Zool., no. 6. 64 pp.

Healy, E. A., & G. P. Wells. 1959. Three new lugworms (Arenicolidae, Polychaeta) from the north Pacific area. Proc. Zool. Soc. London, 133: 315-35.

Hobson, K. D. 1967. The feeding and ecology of two north Pacific *Abarenicola* species (Arenicolidae, Polychaeta). Biol. Bull., 133:343-54.

------. 1971. Some polychaetes of the superfamily Eunicea from the north Pacific and north Atlantic oceans. Proc. Biol. Soc. Wash., 83:527-44.

Pettibone, M. H. 1953. Some Scale-bearing Polychaetes of Puget Sound and Adjacent Waters. Seattle: University of Washington Press. 89 pp.

Ushakov, P. V. 1955. [Polychaete worms of the far-eastern seas of the U.S.S.R.] Opredeliteli po Faune SSSR, 56. 444 pp. (in Russian). (English translation, 1965. Jerusalem: Israel Program for Scientific Translations.)

"ARCHIANNELIDS"

These evidently do not constitute a natural group. A number of worms previously assigned to the so-called Archiannelida have come to rest in various families of polychaetes. Those that remain simply do not fit comfortably among the polychaetes, and for the time being they are treated separately.

Only a few archiannelids have been reported from our region. A key to these would perhaps lead to the misidentification of undescribed species. Therefore, a key to families is provided, and this is followed by a catalogue of recorded species.

1a Head without tentacles, antennae, or palps — Dinophilidae

1b Head with tentacles, antennae, or palps — 2

2a Head with 1 pair of stiff tentacles; body firm, much like that of a nematode, and without a ventral ciliated band; without setae — Polygordiidae

2b Head with flexible tentacles, antennae, or palps; body rather flexible, and with a ventral ciliated band; with or without setae — 3

3a Body short, with fewer than 15 segments, these with setae; prostomium with a pair of palps and with 3 articulated antennae — Nerillidae

3b Body long, with more than 15 segments, these with or without setae; head with a pair of tentacles, but without 3 articulated prostomial antennae or prostomial palps — 4

4a Without setae — Protodrilidae (in general)

4b With setae — 5

5a Setigerous segments with short, retractile parapodia, each with a bundle of setae — Saccocirridae

5b Setigerous segments without parapodia, but with 2 pairs of setae — Protodrilidae: *Protodriloides*

Family Polygordiidae

None has been recorded, but the family is almost certainly represented.

Family Protodrilidae

Protodriloides chaetifer (Remane, 1926) (intertidal, in coarse sand)

Protodrilus flabelliger Wieser, 1957 (intertidal, in coarse sand)

Family Nerillidae

Nerilla antennata O. Schmidt, 1848 (intertidal, in fine sand)

Nerilla digitata Wieser, 1957 (in subterranean water on sandy beaches, above the high tide line)

Nerilla inopinata Gray, 1968 (intertidal, in shell gravel)

Family Saccocirridae

Saccocirrus eroticus Gray, 1969 (intertidal, in shell gravel)

Family Dinophilidae

Trilobodrilus nipponicus Uchida & Okuda, 1943 (intertidal, in sand or shell gravel)

Diurodrilus ankeli Ax, 1967 (intertidal, in coarse sand-gravel)

Dinophilus kincaidi Jones & Ferguson, 1957 (in sediment in a substrate of mud and shells)

Ax, P. 1967. *Diurodrilus ankeli* nov. spec. (Archiannelida) von der nordamerikanischen Pazifikküste. Ein Beitrag zur Morphologie, Systematik und Verbreitung der Gattung *Diurodrilus*. Zeitschr. Morph. Ökol. Tiere, 60:5-16

Gray, J. S. 1968. *Nerilla inopinata,* a new species of archiannelid, from the west coast of North America. Cah. Biol. Mar., 9:441-8.

------. 1969. A new species of *Saccocirrus* (Archiannelida) from the west coast of North America. Pacific Sci., 23: 238-51.

Jones, E. R., & F. F. Ferguson. 1957. The genus *Dinophilus* (Archiannelida) in the United States. Amer. Midl. Nat., 57:440-9.

Wieser, W. 1957. Archiannelids from the intertidal of Puget Sound. Trans. Amer. Micr. Soc., 76:275-85.

CLASS OLIGOCHAETA

Marine representatives of the Oligochaeta are regularly encountered in our region, especially in sediments and under decaying vegetation on the shore. Although several enchytraeids have been described, it is distressing that some common species of other families have not been identified even to genus. The fauna of freshwater and terrestrial oligochaetes of this region is considerably better known, thanks mostly to the comprehensive study of Altman (1936).

Enchytraeus pugetensis Altman, 1931 is a small whitish enchytraeid that is often extremely abundant under mats of seaweed and other material accumulating and decomposing on the shore. *E. multiannulatus,* Altman, 1936 is found under decaying plant material at the edges of salt marshes. More recently, Tynen (1969) described *E. cryptosetosus* and 5 enchytraeids belonging to the genus *Lumbricillus;* all of these occur under decaying seaweed in various habitats.

A paper by Brinkhurst & Simmons (1968) deals with several species, mostly of the family Tubificidae, found in San Francisco Bay; perhaps certain of these should be expected in our area. The review of aquatic oligochaetes by Brinkhurst & Jamieson (1971) characterizes most of the genera and many of the species known to occur in marine habitats. The bibliography in this work will lead the interested student to essential references on marine oligochaetes of various parts of the world.

Altman, L. C. 1931. *Enchytraeus pugetensis*, a new marine enchytraeid from Puget Sound. Trans. Amer. Micr. Soc., 50:154-63.

------. 1936. Oligochaeta of Washington. Univ. Wash. Publ. Biol., 4 (1): 1-137.

Brinkhurst, R. O., & B. G. M. Jamieson. 1971. Aquatic Oligochaeta of the World. Toronto: Univ. of Toronto Press. xi + 860 pp.

Brinkhurst, R. O., & M. L. Simmons. 1968. The aquatic Oligochaeta of San Francisco Bay. Calif. Fish and Game, 54:180-94.

Tynen, M. J. 1969. New Enchytraeidae (Oligochaeta) from the east coast of Vancouver Island. Canad. J. Zool., 47:387-93.

CLASS HIRUDINEA

The marine leeches of the Pacific coast are poorly known. In our area, a few species have been found on fishes (including elasmobranchs), at least 1 occurs on shrimps of the genus *Pandalus*, and 1 is associated with a deep-water sea urchin, *Allocentrotus fragilis*. The references provided may be of some help in identifying certain of these leeches, at least to genus.

Knight-Jones, E. W. 1962. The systematics of marine leeches. *In* Mann, K. H., Leeches (Hirudinea), Their Structure, Physiology, Ecology, and Embryology, pp. 169-186. London: Pergamon Press. x + 201 pp.

Meyer, M. C., & A. A. Barden, Jr. 1955. Leeches symbiotic on Arthropoda, especially decapod Crustacea. Wasmann J. Biol., 13:297-311.

Moore, J. P. 1952. New Piscicolidae (leeches) from the Pacific and their anatomy. Occ. Pap. Bishop Mus., 21:17-44.

Moore, J. P., & M. C. Meyer. 1951. Leeches (Hirudinea) from Alaskan and adjacent waters. Wasmann J. Biol., 9:11-77.

Soós, Á. 1965. Identification key to the leech (Hirudinea) genera of the world, with a catalogue of the species. I. Family: Piscicolidae. Acta Zool., 11:417-63.

PHYLUM PRIAPULIDA

Of the 8 described species of priapulids, *Priapulus caudatus* Lamarck, 1816 is the only one reported from our area. It reaches a length (exclusive of the caudal appendage) of about 5 cm, and has a wide distribution in the Northern Hemisphere. *P. caudatus* is subtidal, and is most easily collected by epibenthic dredging on muddy substrates. The distinctive larva of this species is only a few millimeters long and is probably overlooked much more often than it is recognized.

Hyman, L. H. 1951. The Invertebrates: Acanthocephala, Aschelminthes, Entoprocta. New York: McGraw-Hill Book Co. vii + 572 pp.

Lang, K. 1948. Contribution to the ecology of *Priapulus caudatus* Lam. Arkiv för Zoologi, 41A (5):1-12.

------. 1948. On the morphology of the larva of *Priapulus caudatus* Lam. Arkiv för Zoologi, 41A (9):1-8.

PHYLUM SIPUNCULA

1a Tentacles inconspicuous, fingerlike, not branching 2

1b Tentacles conspicuous when extended, branching dichotomously and appearing bushy *Themiste pyroides* (Chamberlin, 1919)

2a Introvert with dark spots and dark transverse streaks, and with small hooks on its anterior portion; tentacles arranged in a crescentic pattern dorsal to the mouth *Phascolosoma agassizii* Keferstein, 1867

2b Introvert without dark spots or streaks, and without hooks; tentacles arranged in 1 or 2 circles around the mouth 3

3a Introvert about half or slightly more than half as long as the body as a whole; skin nearly smooth (papillae present, but inconspicuous), whitish to dark gray; with 4 retractor muscles *Golfingia pugettensis* Fisher, 1952

3b Introvert less than half as long as the body as a whole; skin decidedly rough, dark brown; with 2 retractor muscles *Golfingia vulgaris* (de Blainville, 1827)

Specimens referred to as *Golfingia vulgaris* lack the small spines characteristic of the introvert of European *G. vulgaris*.

Golfingia margaritacea (Sars, 1851) has been reported from our region (Rat Island, British Columbia; Friday Harbor, Washington) by Fisher (1952). However, no specimens that could be referred to *G. margaritacea* have been taken at Friday Harbor in recent years.

Fisher, W. K. 1952. The sipunculid worms of California and Baja California. Proc. U. S. Nat. Mus., 102:371-450.

Rice, M. E. 1967. A comparative study of the development of *Phascolosoma agassizii*, *Golfingia pugettensis*, and *Themiste pyroides* with a discussion of developmental patterns in the Sipuncula. Ophelia, 4:143-71.

Stephen, A. C. 1964. A revision of the classification of the phylum Sipuncula. Ann. Mag. Nat. Hist., ser. 13, 7:457-62.

Stephen, A. C., & S. J. Edmonds. 1972. The phyla Sipuncula and Echiura. London: British Museum (Natural History). viii + 528 pp.

PHYLUM ECHIURA

1a With 2 nearly complete rings of bristlelike setae at the posterior end of the body *Echiurus echiurus alaskanus* Fisher, 1946

1b Without setae at the posterior end of the body 2

2a With a pair of well-developed setae near the anterior end of the trunk; proboscis (if still attached) troughlike, unbranched; general coloration pinkish to reddish *Arhynchite pugettensis* Fisher, 1949

2b Without setae near the anterior end of the trunk or elsewhere; proboscis (if still attached) bifurcated; dorsal side of trunk greenish, proboscis and ventral side of trunk whitish *Nellobia eusoma* Fisher, 1946

When the genus *Arhynchite* was established, it was characterized by the apparent absence of a proboscis. However, *A. pugettensis* and *A. californicus*, referred by Fisher to this genus on the basis of other morphological features, do possess this structure. The proboscis is easily detached, and incomplete worms may appear perfectly normal. It is possible that even the type species of *Arhynchite* may eventually be shown to have a proboscis.

Fisher, W. K. 1946. Echiuroid worms of the North Pacific Ocean. Proc. U. S. Nat. Mus., 96:215-92.

------. 1949. Additions to the echiuroid fauna of the North Pacific Ocean. Proc. U. S. Nat. Mus., 99:479-97.

Stephen, A. C., & S. J. Edmonds. 1972. The phyla Sipuncula and Echiura. London: British Museum (Natural History). viii + 528 pp.

PHYLUM TARDIGRADA

In this region, tardigrades are fairly common in growths of most mosses and certain lichens. The easiest way to find them is to soak some moss in a small dish of water and, after a few minutes, search the bottom of the dish with a dissecting microscope. The contribution of Schuster & Grigarick (1965) covers the species likely to be encountered.

Echiniscoides sigismundi (M. Schultze, 1865) is a marine tardigrade with a wide distribution, and has been found at various points along the Pacific coast, including some localities in our area. It appears to be most common among the filamentous green algae that grow on barnacles and mussels, as well as in a similar situation on rocks. In sandy beaches, interstitial species such as *Batillipes mirus* Richters, 1909 should be expected.

Crisp, D. J., & J. Hobart. 1954. A note on the habitat of the marine tardigrade *Echiniscoides sigismundi* (Schultze). Ann. Mag. Nat. Hist., ser. 12, 7:554-60.

Marcus, E. 1929. Tardigrada. *In* Bronn, H. G., Klassen und Ordnungen des Tierreichs, V: IV: 3. vii + 608 pp.

Renaud-Mornant, J., & L. W. Pollock. 1971. A review of the systematics and ecology of marine Tardigrada. Smithsonian Contr. Zool., no. 76:106-17.

Schulz, E. 1955. Studien an marinen Tardigraden. Kieler Meeresforsch., 11: 73-9.

Schuster, R. O., & A. A. Grigarick. 1965. Tardigrada from western North America, with emphasis on the fauna of California. Univ. Calif. Publ. Zool., 76:1-67.

PHYLUM ARTHROPODA

SUBPHYLUM CHELICERATA

CLASS PYCNOGONIDA

Joel W. Hedgpeth
Marine Science Center, Oregon State University, Newport

The key is based on characters of adult pycnogonids. Differences between the sexes are avoided as much as possible. It must be remembered, however, that these animals are sexually dimorphic, and that the males carry the eggs. Mature specimens without eggs may be females, but not necessarily so, unless they also lack ovigers. As pycnogonids do not have the characteristic complement of segments and appendages until they are fully adult, it may be impossible to assign juveniles to genus or even to family. One should try to see as many specimens as possible from a particular situation, with the hope that some of them will be mature. Young specimens are often found in hydroids, molluscs, or other soft-bodied invertebrates.

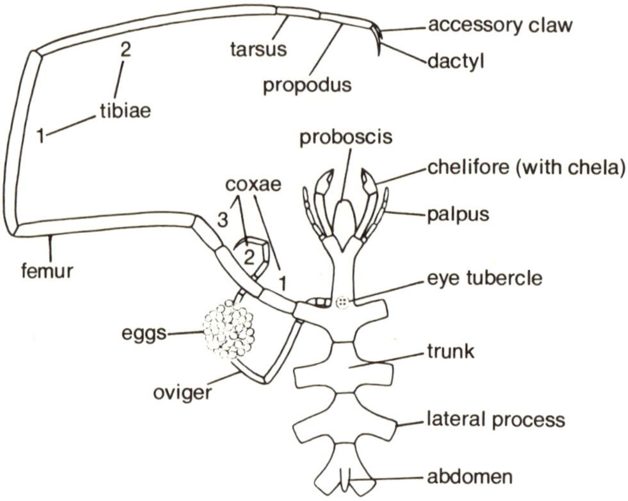

1a Legs thick, shorter than (or at least not conspicuously longer than) the combined length of the proboscis and trunk; outline of the animal as a whole oval or elliptical; without chelifores or palpi; without conspicuous spiny processes 2

1b Legs usually much longer than the combined length of the proboscis and trunk (pycnogonids of this type are usually awkward and gangly, but sometimes are fairly compact); outline of the animal as a whole approximately circular; with either chelifores or palpi, or both; often spiny or setose 3

2a Color uniform (ivory to pink); height of dorsal tubercles, when these are present, not greater than their basal diameter *Pycnogonum stearnsi* Ives, 1892

2b Color not uniform (generally light brown with dark lines, or with clear areas between dark patches); height of dorsal tubercles greater than their basal diameter *Pycnogonum rickettsi* Schmitt, 1934
In both of these species of *Pycnogonum*, adult males are smaller than adult females. The largest females lack ovigers.

3a With both chelifores and palpi (the palpi are conspicuous and have a number of articles); chelifores either with functional chelae or with chelae reduced to knobs, or achelate and reduced to stubs (however, juveniles of some species have chelae that are lost as the animals mature); ovigers usually present in both sexes and conspicuous 4

3b With chelifores, but without palpi; chelifores with chelae and conspicuous; ovigers reduced or entirely absent in females 17

4a Chelifores with 2 basal articles; small chelae sometimes present (however, the chelae may be reduced to non-functional knobs) 5

4b Chelifores with a single basal article (scape); with a chela or a knoblike second article (or the entire appendage may be reduced to a knob or to an elongate, tuberclelike structure) 7

5a Very spiny in appearance, with prominent double rows of tall spiny processes on the longer articles of the legs and a single median row of spiny processes on the dorsum of the trunk *Nymphopsis spinosissima* (Hall, 1912)

5b Smooth or spiny, but if spiny, without tall processes arranged as described in choice 5a 6

6a Trunk compact, circular; legs without long spines; with 2 or 3 tubercles along the median part of the dorsum of the trunk *Ammothella tuberculata* Cole, 1904

6b Trunk not so compact as to be circular; legs with long spines, especially on the longer articles; without tubercles along the median part of the dorsum of the trunk *Ammothella* sp.

7a Chelifores reduced to knobs or tubercles 8

7b Chelifores consisting of a basal article (scape) and either a well developed terminal chela that extends beyond the proboscis or a knoblike article (if functional chelae are present in sub-adults, these do not extend beyond the proboscis) 11

8a Each chelifore consisting of an easily recognized, papillate process; a gangly species with rather long and slender legs, these often with brownish purple bands; palpi with 9 articles *Lecythorhynchus hilgendorfi* (Böhm, 1879)

8b Chelifores small and difficult to see; animal rather compact; palpi with 6 or fewer articles 9

9a Body covered with short setae, thus appearing hairy; associated with the anthozoan *Gersemia* *Tanystylum anthomasti* Hedgpeth, 1949

9b Body not covered with short setae (however, it may have sparse spines); not normally associated with *Gersemia* 10

10a Proboscis about twice the length of basal diameter, rounded at the tip; trunk very compact, the lateral processes indistinct *Tanystylum occidentalis* (Cole, 1904)

10b Proboscis at least 3 times as long as basal diameter, tapering to a blunt, subconical tip; trunk not so compact as in *T. occidentalis*, the lateral processes touching, but distinct *Tanystylum* sp.

11a Chelifores conspicuous, chelate, extending beyond the proboscis; palpi slender, with 5 articles; total extent with legs outstretched sometimes exceeding 5 cm; usually subtidal 12

11b Chelifores usually reduced to knobs in the adult, but if chelae persist, the chelifores are shorter than the proboscis; palpi with 8 articles; total extent less than 2 cm; includes some intertidal species 13

12a Chelae slender, the dactyl and fixed finger about as long as the palm, and gracefully curved; tarsus of legs as long as, or longer than, the propodus; legs slender *Nymphon pixellae* Scott, 1913

12b Chelae stout, the dactyl and fixed finger less than half as long as the palm; tarsus of legs usually shorter than the propodus; legs stout, sparsely spinose (a highly variable species) *Nymphon grossipes* (Fabricius, 1780)

13a Adults not chelate 14

13b Adults chelate, the dactyl and fixed finger arcuate (a parasite of mussels in central California; found at least as far north as Boiler Bay, Oregon, but the habitat on the Oregon coast is not known) *Achelia chelata* (Hilton, 1939)

14a Dorsodistal edges of first coxal articles with fingerlike processes or short outgrowths 15
14b Dorsodistal edges of first coxal articles without fingerlike processes or short outgrowths (however, there may be low knobs or wartlike eminences on the lateral processes of the trunk) *Achelia latifrons* (Cole, 1904)
15a Processes on first coxal articles less than half as long as the diameter of the articles; proboscis broadly ellipsoidal (usually in bays) *Achelia nudiuscula* (Hall, 1913)
15b Processes on first coxal articles at least half as long as the diameter of the articles; proboscis narrowly ellipsoidal 16
16a Processes on first coxal articles at least three-fourths as long as the diameter of the articles; eye tubercle erect *Achelia gracilipes* (Cole, 1904)
16b Processes on first coxal articles about half as long as the diameter of the articles; eye tubercle low and broad *Achelia alaskensis* (Cole, 1904)
17a Lateral processes very close together; auxiliary claws minute and inconspicuous; legs slender 18
17b Lateral processes separated by at least half their diameter; auxiliary claws small, but well developed; legs stout *Phoxichilidium femoratum* (Rathke, 1799)
18a Intestine bright green, its branches clearly visible in the legs; lateral processes close together, but not quite touching; eye tubercle at least twice as high as broad *Halosoma viridintestinale* Cole, 1904
18b Intestine not bright green; lateral processes touching one another; eye tubercle broader than high *Halosoma compactum* Hilton, 1939

Cole, L. J. 1904. Pycnogonida of the west coast of North America. Harriman Alaska Expedition, 10:249-98.

Exline, H. I. 1936. Pycnogonida from Puget Sound. Proc. U. S. Nat. Mus., 83:413-22.

Hedgpeth, J. W. 1941. A key to the Pycnogonida of the Pacific coast of North America. Trans. San Diego Soc. Nat. Hist., 9:253-64.

------. 1951. Pycnogonids from Dillon Beach and vicinity, California, with descriptions of two new species. Wasmann J. Biol., 9:105-117.

Scott, F. M. 1913. On a species of *Nymphon* from the north Pacific. Ann. Mag. Nat. Hist., ser. 8, 10:206-9.

CLASS ARACHNIDA

ORDER PSEUDOSCORPIONIDA (CHELONETHIDA)

The only distinctly maritime species reported from this area is *Halobisium occidentale* Beier, 1931. Its maximum length is about 4 mm. It is most likely to be found at the edges of *Salicornia* marshes, under rocks or debris just above the high tide line.

Chamberlin, J. C. 1931. The arachnid order Chelonethida. Stanford Univ. Publ. Biol. Sci., 7:1-284.

ORDER ACARI

The three major suborders of mites--Parasitiformes, Trombidiformes, and Sarcoptiformes--have marine representatives. However, just 1 family of the Trombidiformes, the Halacaridae, includes most of the genera likely to be encountered in marine habitats. The halacarids almost certainly do not constitute a natural assemblage; they seem to have evolved from several groups of terrestrial mites. In any case, about 15 of the genera assigned to this family are reported for the northeastern Pacific region.

The key prepared by Newell for *Light's Manual* covers the genera of Halacaridae occuring in our area. It does not deal with species, of which there must be many, mostly still undescribed. Anyone contemplating work on halacarids will find the various papers of Newell·indispensable; those most likely to be useful here are listed in the bibliography.

A mite that frequently parasitizes the amphipod *Orchestoidea californiana*, found in the drift zone on exposed sandy beaches, was described by Canaris (1962) as *Gammaridacarus brevisternalis*. It is assigned to the suborder Parasitiformes.

Canaris, A. G. 1962. A new genus and species of mite (Laelaptidae) from *Orchestoidea californiana* (Gammaridea). J. Parasit., 48:467-9.

Newell, I. M. 1947. A systematic and ecological study of the Halacaridae of eastern North America. Bull. Bingham Oceanogr. Coll., 10(3):1-232.

------. 1949. New genera and species of Halacaridae (Acari). Amer. Mus. Novit., no. 1411. 22 pp.

------. 1950. New species of *Copidognathus* from the Aleutians. Amer. Mus. Novit., no. 1476. 19 pp.

------. 1951. New species of *Agaue* and *Thalassarachna* from the Aleutians. Amer. Mus. Novit., no. 1489. 19 pp.

------. 1952. *Copidognathus curtus* Hall, 1912, and other species of *Copidognathus* from western North America. Amer. Mus. Novit., no. 1499. 27 pp.

------. 1952. Further studies on Alaskan Halacaridae. Amer. Mus. Novit., no. 1536. 56 pp.

------. 1953. The natural classification of the Rhombognathinae (Acari, Halacaridae). Syst. Zool., 2:119-35.

------. 1971. Halacaridae (Acari) collected during Cruise 17 of the R/V ANTON BRUUN in the southeastern Pacific Ocean. ANTON BRUUN Report Number 8 (Scientific Results of the Southeast Pacific Expedition):1-58.

SUBPHYLUM MANDIBULATA

CLASS CRUSTACEA

SUBCLASS BRANCHIOPODA

ORDER CLADOCERA

As far as is known, only 2 genera of marine cladocerans, both members of the family Polyphemidae, occur in this region. They can be differentiated on the basis of characters summarized in the key. The references listed in the bibliography will be helpful to those who need to identify the species.

1a Head separated from the rest of the body by a deep groove; dorsal pouch rounded *Podon*
(*P. polyphemoides* (Leuckart, 1859) appears to be the common species in our area; however, *P. leuckarti* (G. O. Sars, 1861) should probably be expected.)

1b Head not separated from the rest of the body by a groove; dorsal pouch typically at least somewhat pointed, so that the body as a whole (as seen in side view) is nearly triangular *Evadne*
(*E. nordmanni* Lovén, 1835 is the prevailing species. However, *E. tergestina* Claus, 1877 has been reported, and *E. spinifera* P. E. Müller, 1868 perhaps also occurs.)

Apstein, C. 1901. Cladocera (Daphnidae), Wasserflöhe. Nordisches Plankton, Zool., 4, Abt. VII:11-15.

------. 1911. Die Cladoceren (Daphniden). Nachtrag. Nordisches Plankton, Zool., 4, Abt. VII:17-20.

Baker, H. M. 1938. Studies on the Cladocera of Monterey Bay. Proc. Calif. Acad. Sci., ser. 4, 23:311-65.

Rammner, W. 1930. Phyllopoda. *In* Grimpe, G., and E. Wagler, Die Tierwelt der Nord- und Ostsee, 10a. 32 pp.

------. 1939. Cladocera. Fiches d'Identification du Zooplancton (Conseil Permanent International pour l'Exploration de la Mer), No. 3.

SUBCLASS OSTRACODA

There are no detailed works on the ostracods of our region. The papers of Lucas (1930) and Smith (1952) recorded the occurrence of 35 species, mostly near Nanaimo, on Vancouver Island, and Wieser (1959) reported 7 species from Puget Sound. (None of the species listed by Wieser was among those found by Lucas and Smith, but perhaps nomenclatorial changes or differences in interpretation of species descriptions partially account for the apparent lack of concordance.)

Under the circumstances, it seems best to refer those who would identify ostracods from this area to the papers cited above, and also to the work of Skogsberg, which treats species of *Cythereis* in detail. The contributions of Elofson (1941), Juday (1907), Klie (1929), and Sars (1922-28) will be helpful in recognition of genera (and perhaps certain species) that may be encountered.

An ostracod associated with the amphipod *Amphithoë humeralis* has been described as *Acetabulosoma kozloffi* by Hart (1971). It is attached to the underside of the thoracic region of its host.

Elofson, O. 1941. Zur Kenntnis der marinen Ostracoden Schwedens. Zool. Bidr. Uppsala, 19:215-534.

Hart, C. W., Jr. 1971. A new species of parasitic ostracod of the genus *Acetabulosoma* (Paradoxostomatidae, Paradoxostominae) with a discussion of the copulatory appendage homologies. Notulae Naturae Acad. Nat. Sci. Philadelphia, no. 442. 11 pp.

Juday, C. 1907. Ostracoda of the San Diego region. II. Littoral forms. Univ. Calif. Publ. Zool., 3:135-56.

Klie, W. 1929. Ostracoda. *In* Grimpe, G., and E. Wagler, Die Tierwelt der Nord- und Ostsee, Teil 10b. 56 pp.

Lucas, V. Z. 1931. Ostracoda of the Vancouver Island region. Contrib. Canad. Biol. Fish., n. s., 6:397-416.

McHardy, R. A. 1964. Marine ostracods from the plankton of Indian Arm, British Columbia, including a diminutive subspecies resembling *Conchoecia alata major* Rudjakov. J. Fish. Res. Bd. Canada, 21:555-76.

Neale, J. W. (ed.). 1969. The Taxonomy, Morphology, and Ecology of Recent Ostracoda. Edinburgh: Oliver & Boyd. ix + 553 pp.

Sars, G. O. 1922-28. Ostracoda. *In* An Account of the Crustacea of Norway, vol. 9. 277 pp.

Skogsberg, T. 1928. Studies on marine ostracods, Part II. External morphology of the genus *Cythereis* with descriptions of twenty-one new species. Occ. Pap. Calif. Acad. Sci., no. 15. 155 pp.

Smith, V. Z. 1952. Further Ostracoda of the Vancouver Island region. J. Fish. Res. Bd. Canada, 9:16-41.

Watling, L. 1970. Two new species of Cytherinae from central California. Crustaceana, 19:251-63.

------. 1972. A new species of *Acetabulastoma* Skornikov from central California with a review of the genus. Proc. Biol. Soc. Wash., 85:481-8.

Wieser, W. 1959. Free-living nematodes and other small invertebrates of Puget Sound beaches. Univ. Wash. Publ. Biol., 19. xi + 179 pp.

SUBCLASS COPEPODA

ORDER CALANOIDA

The primary reference on calanoid copepods for this region is the work of Davis (1949). It includes a key to genera and descriptions of about 60 species. The key provided by Fulton (1968) does not cover as many species, but is practical in the sense that it deals with calanoids regularly known to occur in the Strait of Georgia. The unorthodox form of the key makes it difficult to use, but there are helpful illustrations to keep one on the right track. A substantially complete checklist in Fulton's work is valuable for indicating the approximate depth and the season at which each species is found, and for citing published references to occurrence of each species in our general area.

The bibliography of Davis thoroughly covers pertinent earlier works; that of Fulton lists a number of recent contributions that will be important to anyone concerned with calanoids in this region. The bibliography below includes only a few papers that are especially useful because they are comprehensive, are based on surveys of copepods of this region, deal with recently described species, or will lead to other references.

Brodskiĭ, K. A. 1950. [Calanoida of the far-eastern seas of the U.S.S.R. and of the Polar basin.] Opredeliteli po Faune SSSR, 35. 442 pp. (in Russian)

Campbell, M. H. 1929. Some free-swimming copepods of the Vancouver Island region. Trans. Royal Soc. Canada, Sect. V, ser. 3, 23:303-32.

------. 1930. Some free-swimming copepods of the Vancouver Island region. II. Trans. Royal Soc. Canada, Sect. V, ser. 3, 24:177-84.

Damkaer, D. M. 1971. *Parastephos occatum*, a new species of hyperbenthic copepod (Calanoida: Stephidae) from the inland marine water of Washington State. Proc. Biol. Soc. Washington, 83:505-14.

Davis, C. C. 1949. The pelagic Copepoda of the northeastern Pacific Ocean. Univ. Wash. Publ. Biol., 14. iv + 117 pp.

Esterly, C. O. 1924. The free-swimming Copepoda of San Francisco Bay. Univ. Calif. Publ. Zool., 26:81-129.

Frost, B. W. 1971. Taxonomic status of *Calanus finmarchicus* and *C. glacialis* (Copepoda), with special reference to adult males. J. Fish. Res. Bd. Canada, 28:23-30.
Frost, B., & A. Fleminger. 1968. A revision of the genus *Clausocalanus* (Copepoda: Calanoida) with remarks on distributional patterns in diagnostic characters. Bull. Scripps Inst. Oceanogr., 12:1-99.
Fulton, J. 1968. A Laboratory Manual for the Identification of British Columbia Marine Zooplankton. Technical Rep. no. 55, Fisheries Research Board of Canada. iii + 141 pp.
Johnson, M. W. 1932. Seasonal distribution of plankton at Friday Harbor, Washington. Univ. Wash. Publ. Oceanogr., 1:1-38.
Park, T. S. 1966. A new species of *Bradyidius* (Copepoda: Calanoida) from the Pacific coast of North America. J. Fish. Res. Bd. Canada, 23:805-11.
------. 1967. Two new species of calanoid copepods from the Strait of Georgia, British Columbia, Canada. J. Fish. Res. Bd. Canada, 24:231-42.
Wilson, C. B. 1932. The copepods of the Woods Hole region, Massachusetts. Bull. 158, U. S. Nat. Mus. xix + 635 pp.

ORDER HARPACTICOIDA

The most important single work on marine harpacticoids of the Pacific coast is that of Lang (1965). Although Lang was concerned primarily with species found in central California, he attempted to evaluate the lists of harpacticoids reported for other regions of the Pacific coast. Of the nearly 100 species he recorded, about 80 were new. Chappuis (1958), on the other hand, listed 38 species for Puget Sound (many of them referred to also by Wieser, 1959), only 3 of which were new. Lang suggested that Chappuis had given existing names to many species that in reality were new.

The papers listed in the bibliography are the more comprehensive ones dealing with Pacific coast harpacticoids. They will lead to most of the minor contributions that the specialist may need to consult. Attention is also called to the study of Fahrenbach (1962) which deals principally with *Diarthrodes cystoecus*, found in close association with certain red algae; his work also briefly mentions *Thalestris rhodymeniae*, which inhabits the bladders of the red alga *Halosaccion glandiforme* (in California, this species harbors *D. cystoecus*).

Campbell, M. H. 1929. Some free-swimming copepods of the Vancouver Island region. Trans. Royal Soc. Canada, Sect. V, ser. 3, 23:303-32.
------. 1930. Some free-swimming copepods of the Vancouver Island region. II. Trans. Royal Soc. Canada, Sect. V, ser. 3, 24:177-84.
Chappuis, P. A. 1958. Harpacticoïdes psammiques marins des environs de Seattle (Washington, U. S. A.). Vie et Milieu, 8 (1957):409-22.
Fahrenbach, W. I. 1962. The biology of a harpacticoid copepod. La Cellule, 62:301-76.
Lang, K. 1948. Monographie der Harpacticiden. 2 vols. Stockholm: Haken Ohlssons Boktrycheri. 1683 pp.
------. 1965. Copepoda Harpacticoidea from the California Pacific coast. Kungl. Svenska Vetenskapakad. Handl., ser. 4, 10, no. 2. 560 pp.
Monk, C. R. 1941. Marine harpacticoid copepods from California. Trans. Amer. Micr. Soc., 60:75-99.
Wieser, W. 1959. Free-living nematodes and other small invertebrates of

Puget Sound beaches. Univ. Wash. Publ. Biol., 19. xi + 179 pp.

ORDER CYCLOPOIDA

Free-Swimming Species

Only a few species have been reported from our area. According to Davis (1949), these belong to the genera *Oithona, Corycaeus, Oncaea,* and *Conaea.*

Symbiotic Species

These are associated with a wide variety of invertebrate hosts. The classification of symbiotic cyclopoids is so involved that no attempt will be made here to categorize the families or other taxa. Instead, attention will be called to particular species that are likely to be encountered and to the literature dealing with symbiotic cyclopoids occurring in this region. The comprehensive study of Humes & Stock (1973) of 5 families of the superfamily Lichomolgoidea provides an extremely useful summary of this general group, on a world-wide basis.

Family Clausidiidae

Clausidium vancouverense (Haddon, 1912), a reddish species regularly found in the branchial chamber of the ghost shrimps, *Callianassa californiensis* and *C. gigas,* and the mud shrimp, *Upogebia pugettensis* (see Light & Hartman, 1937)

Hemicyclops thysanotus Wilson, 1935 *(=H. pugettensis* Light & Hartman, 1937), occurring with *Clausidium vancouverense* in the branchial chamber of *Callianassa gigas* (farther south, it has been taken also on *C. californiensis, Upogebia pugettensis,* and the nudibranch *Hermissenda crassicornis)*

Hemicyclops subadhaerens Gooding, 1960: in burrows of *Callianassa californiensis*

Family Lichomolgidae

Doridicola ptilosarci Humes & Stock, 1973, between the flaps and on the surface of the stem of the sea pen, *Ptilosarcus gurneyi.*

Family Pseudanthessiidae

Pseudanthessius latus Illg, 1950, commensal with a marine polyclad turbellarian, *Kaburakia excelsa (Cryptophallus magnus)*

Family Sabelliphilidae

Several species of *Hermannella,* on the ctenidia of various bivalve molluscs; mostly described by Illg (1949) as species of *Paranthessius* (see Humes & Stock, 1973)

Family Myicolidae

Anthessius nortoni Illg, 1960, associated with the keyhole limpet, *Diodora aspera*

Family Gastrodelphyidae

Several species of *Gastrodelphys* and *Sabellacheres* associated with sabellid polychaete annelids (Dudley, 1964)

Family Notodelphyidae

A number of species associated with ascidians (Illg, 1958; Dudley, 1966)

Davis, C. C. 1949. The pelagic Copepoda of the northeastern Pacific Ocean. Univ. Wash. Publ. Biol., 14. iv + 117 pp.

Dudley, P. L. 1964. Some gastrodelphyid copepods from the Pacific coast of North America. Amer. Mus. Novit., no. 2194. 51 pp.

------. 1966. Development and systematics of some Pacific marine symbiotic copepods. A study of the biology of the Notodelphyidae, associates of ascidians. Univ. Wash. Publ. Biol., 21. 282 pp.

Gooding, R. V. 1960. North and South American copepods of the genus *Hemicyclops* (Cyclopoida: Clausidiidae). Proc. U. S. Nat. Mus., 112:159-95.

Humes, A. G., & J. H. Stock. 1973. A revision of the family Lichomolgidae Kossmann, 1877, cyclopoid copepods mainly associated with marine invertebrates. Smithsonian Contr. Zool., no. 127. v + 368 pp.

Illg, P. L. 1949. A review of the copepod genus *Paranthessius* Claus. Proc. U. S. Nat. Mus., 99:391-428.

------. 1950. A new copepod, *Pseudanthessius latus* (Cyclopoida: Lichomolgidae), commensal with a marine flatworm. J. Wash. Acad. Sci., 40:129-33.

------. 1958. North American copepods of the family Notodelphyidae. Proc. U. S. Nat. Mus., 107:463-649.

------. 1960. Marine copepods of the genus *Anthessius* from the northeastern Pacific Ocean. Pacific Sci., 14:337-72.

Light, S. F., & O. Hartman. 1937. A review of the genera *Clausidium* Kossman and *Hemicyclops* Boeck (Copepoda, Cyclopoida). Univ. Calif. Publ. Zool., 41:173-87.

ORDER MONSTRILLOIDA

The members of this group are parasitic (largely in polychaete annelids) during most of their life history. When they reach maturity, they emerge and become free-swimming. The adults lack mouthparts and cannot feed, so must subsist on reserves accumulated during the parasitic phase; they live only for a short time.

Altogether, 4 species have been reported from this general area: *Monstrilla longiremis* Giesbrecht, 1892 (see Wailes, 1929); *M. wandelii* Stephensen, 1913; *M. helgolandica* Claus, 1863; *M. spinosa* Park, 1967. The last 3 were carefully described by Park (1967). Davis (1949) provided a review of the Monstrilloida in general.

Davis, C. C. 1949. A preliminary revision of the Monstrilloida, with descriptions of two new species. Trans. Amer. Micr. Soc., 68:245-55.

Park, T. S. 1967. Two unreported species and one new species of *Monstrilla* (Copepoda: Monstrilloida) from the Strait of Georgia. Trans. Amer. Micr. Soc., 86:144-52.

Wailes, G. H. 1929. The marine zooplankton of British Columbia. Art Hist. Sci. Assoc., Vancouver, B. C., Mus. Art Notes, 4:159-69.

ORDER CALIGOIDA: ORDER LERNAEOPODOIDA

These are parasites of fishes, and many species have been reported from this region. Yamaguti (1963) has provided a catalogue and a guide to the literature, essentially cosmopolitan in scope, for both groups. The numerous papers of Wilson may be helpful in many respects; the list below is not complete. A very productive worker in our area at the present time is Kabata. His contributions will lead other specialists to most of the essential literature. Only a few of Kabata's more comprehensive recent papers dealing with caligoids and lernaeopodoids are cited.

Fraser, C. McL. 1920. Copepods parasitic on fish from the Vancouver Island Region. Trans. Royal Soc. Canada, Sect. V, ser. 3, 13 (1919):45-67.

Kabata, Z. 1968. Some Chondracanthidae (Copepoda) from fishes of British Columbia. J. Fish. Res. Bd. Canada, 25:321-45.

------ 1969. Revision of the genus *Salmincola* Wilson, 1915 (Copepoda: Lernaeopodidae). J. Fish. Res. Bd. Canada, 26:2987-3041.

------. 1973. The species of *Lepeophtheirus* (Copepoda: Caligidae) from fishes of British Columbia. J. Fish. Res. Bd. Canada, 30:729-59.

Parker, R. R., Z. Kabata, L. Margolis, & M. D. Dean. 1968. A review and description of *Caligus curtus* Müller, 1785 (Caligidae: Copepoda), type species of its genus. J. Fish. Res. Bd. Canada, 25:1923-1969.

Wilson, C. B. 1905. North American parasitic copepods belonging to the family Caligidae. Part 1. Caliginae. Proc. U. S. Nat. Mus., 28:479-672.

------. 1907. North American parasitic copepods belonging to the family Caligidae. Part 2. The Trebinae and Euryphorinae. Proc. U. S. Nat. Mus., 31:669-720.

------. 1907. North American parasitic copepods belonging to the family Caligidae. Parts 3 and 4. A revision of the Pandarinae and the Cecropinae. Proc. U. S. Nat. Mus., 33:323-490.

------. 1912. Parasitic copepods from Nanaimo, British Columbia, including eight species new to science. Contrib. Canada Biol. Fish., 1906-1910: 85-101.

------. 1932. The copepods of the Woods Hole Region, Massachusetts. Bull. 158, U. S. Nat. Mus. xix + 635 pp.

------. 1935. Parasitic Copepoda from the Pacific coast. Amer. Midl. Nat., 16:776-97.

------. 1944. Parasitic copepods in the United States National Museum. Proc. U. S. Nat. Mus., 94:529-82.

Yamaguti, S. 1963. Parasitic Copepoda and Branchiura of Fishes. New York & London: Interscience Publishers. vii + 1104 pp.

SUBCLASS BRANCHIURA

These are ectoparasites of freshwater and marine fishes. There are relatively few species, and those occurring in our region can probably be identified with the aid of the references cited.

Cressey, R. F. 1972. The genus *Argulus* (Crustacea: Branchiura) of the United States. Identification Manual No. 2, Biota of Freshwater Ecosystems, U. S. Environmental Protection Agency. viii + 14 pp.

Meehean, O. L. 1940. A review of the parasitic Crustacea of the genus *Argulus* in the collections of the United States National Museum. Proc. U. S. Nat. Mus., 88:459-522.

Wilson, C. B. 1944. Parasitic copepods in the United States National Museum. Proc. U. S. Nat. Mus., 94:529-82.
Yamaguti, S. 1963. Parasitic Copepoda and Branchiura of Fishes. New York & London: Interscience Publishers. vii + 1104 pp.

SUBCLASS CIRRIPEDIA

ORDER THORACICA
Barnacles Other than Those Found on Whales

1a Attached to the substrate by a somewhat flexible peduncle (Suborder Lepadomorpha 2
1b Without a flexible peduncle, the shell itself being fused to the substrate (Suborder Balanomorpha) 7

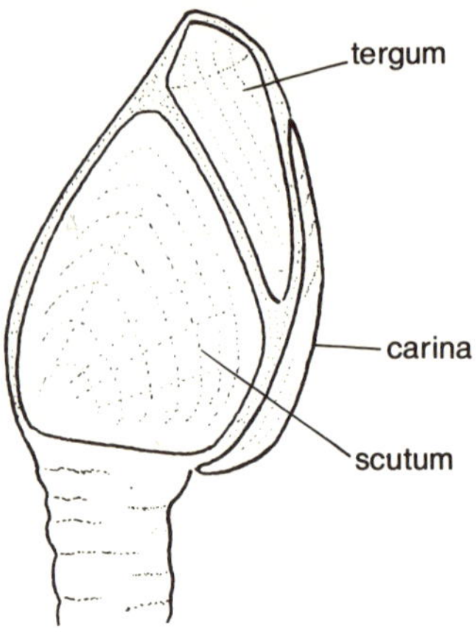

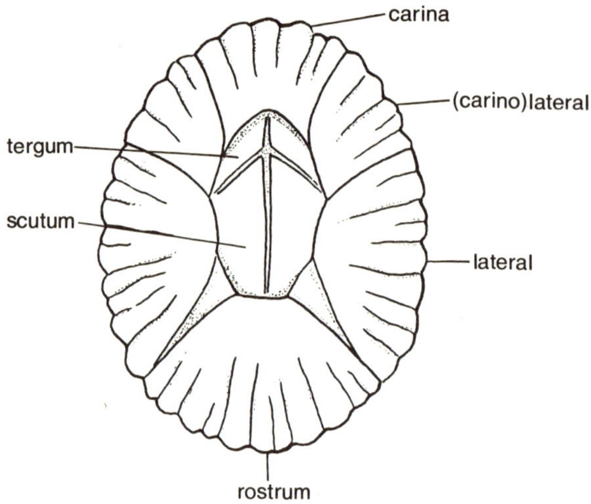

2a Capitulum with 13 or 14 plates; with distinct spines on the ridges of the carina *Scalpellum columbianum* Pilsbry, 1909
2b Capitulum with fewer than, or more than, 13 or 14 plates (when there are more, most of them are small and scalelike); without distinct spines on the ridges of the carina 3
3a Capitulum with 5 large plates and numerous small plates in several rows; in areas with considerable wave action *Pollicipes polymerus* (Sowerby, 1833)
3b Capitulum with only 5 large plates; attached to floating timbers or secreting a float, not attached to rocks 4
4a Capitulum decidedly compressed; attached to floating timbers 5
4b Capitulum nearly globular; associated individuals attached to a common float secreted at the bases of the peduncles *Lepas fascicularis* Ellis & Solander, 1786
5a Tergum without a notch on the side bordering the scutum 6
5b Tergum with a rather distinct notch on the side bordering the scutum *Lepas pacifica* Henry, 1940
6a Plates with fine striations; not more than 2 filamentary appendages at the base of each of the first cirri *Lepas anatifera* (Linnaeus, 1758)
6b Plates smooth; 3 or more filamentary appendages at the base of each of

the first cirri *Lepas hilli* (Leach, 1818)

7a Rostrum overlapping the plates on either side of it; spaces between the cover-plates not in the form of a cross; diameter regularly exceeding 1 cm; at various levels in the intertidal region or subtidal 8

7b Rostrum overlapped by plates on either side of it; spaces between the cover-plates in the form of a cross; diameter under 1 cm; almost completely restricted to the upper intertidal region *Chthamalus dalli* Pilsbry, 1916

8a Base of shell calcified (a calcareous scar remains on the substrate when the barnacle is broken off); ridges of wall, if prominent, not giving rise to downward-pointing, fingerlike excrescences 9

8b Base of shell not calcified (soft tissue remains on the substrate when the barnacle is broken off); ridges of wall prominent, and giving rise, especially in the lower half of the shell, to downward-pointing, fingerlike excrescences (this characteristic may not be evident in crowded or eroded individuals) *Balanus cariosus* (Pallas, 1788)

9a With tubes in the wall (to demonstrate these, the wall will have to be worn away with a fine file or a grindstone) 10

9b Without tubes in the wall 14

10a Without cross-septa in tubes of the wall 11

10b With cross-septa in tubes of the wall (in species that accumulate powder in the tubes, this may have to be blown out before the septa are visible 12

11a Interior of cover-plates white; base of shell solid, with radial grooves; height of shell rarely exceeding 3.5 cm *Balanus balanus pugetensis* Pilsbry, 1916

11b Interior of cover-plates buff; base of shell irregularly porous; diameter of shell commonly attaining 10 cm, sometimes larger *Balanus nubilus* Darwin, 1854

12a Tubes filled with white powder; abundant in upper intertidal region *Balanus glandula* Darwin, 1854

12b Tubes of wall hollow; mostly subtidal and in lower intertidal region 13

13a Cross-septa present only in tubes of upper part of wall; exterior of wall nearly white; diameter rarely exceeding 2.5 cm *Balanus crenatus* Bruguière, 1789

13b Cross-septa present in tubes of wall from base to apex; some portions of the wall, where the epidermis has not been eroded, retaining a brownish color; diameter sometimes attaining 5 cm *Balanus rostratus alaskensis* Pilsbry, 1916

14a Wall rough and strongly ribbed; orifice oval *Balanus engbergi* Pilsbry, 1921

14b Wall smooth and white; orifice nearly square *Balanus hesperius laevidomus* Pilsbry, 1916

Whale Barnacles

The barnacles found on whales are of 2 entirely different types. Those that resemble acorn barnacles of the genera *Balanus* and *Chthamalus* belong in the same suborder, Balanomorpha. Those that resemble goose barnacles are in the suborder Lepadomorpha. The barnacles likely to be found on whales in our region are listed below. All of them are concisely described by Cornwall (1969).

Suborder Lepadomorpha: Family Lepadidae

Conchoderma auritum (Linnaeus, 1758)

attached to shells of *Coronula* on the humpback whale *(Megaptera nodosa)*; sometimes on ship bottoms and floating objects

Conchoderma virgatum (Spengler, 1790)

attached to shells of *Coronula* on the humpback whale *(Megaptera nodosa)*; sometimes on ship bottoms and floating objects, or attached to copepods parasitic on whales or larger fishes

Suborder Balanomorpha: Family Balanidae: Subfamily Coronulinae

Coronula diadema (Linnaeus, 1767)

on the humpback whale *(Megaptera nodosa)*

Coronula reginae Darwin, 1854

on the humpback whale *(Megaptera nodosa)*

Cryptolepas rachianecti Dall, 1872

on the California gray whale *(Eschrichtius glaucus)* (almost wholly embedded in the skin)

Xenobalanus globicipitis Steenstrup, 1851

on the sei whale *(Balaenoptera borealis)*; common finback whale *(Balaenoptera physalus)*; and the blackfish or pilot whale *(Globicephala ventricosa)*

Cornwall, I. E. 1969. The barnacles of British Columbia. 2nd ed. Handbook No. 7, British Columbia Provincial Museum, Victoria. 69 pp.

Henry, D. P. 1940. The Cirripedia of Puget Sound with a key to the species. Univ. Wash. Publ. Oceanogr., 4:1-48.

------. 1940. Notes on some pedunculate barnacles from the North Pacific. Proc. U. S. Nat. Mus., 88:225-36.

------. 1942. Studies on the sessile Cirripedia of the Pacific coast of North America. Univ. Wash. Publ. Oceanogr., 4:95-134.

Pilsbry, H. A. 1907. Cirripedia from the Pacific coast of North America. Bull. U. S. Bur. Fish., 26:193-204.

------. 1916. The sessile barnacles (Cirripedia) contained in the collections of the U. S. National Museum; including a monograph of the American species. Bull. 93, U. S. Nat. Mus. xi + 366 pp.

------. 1921. Barnacles of the San Juan Islands, Washington. Proc. U. S. Nat. Mus., 59:111-15.

ORDER ACROTHORACICA

The barnacles of this group burrow into shells of other barnacles, as well as into shells of molluscs and calcareous skeletons of corals. One species, *Trypetesa lateralis* Tomlinson, 1953, is abundant in central California in gastropod shells inhabited by hermit crabs, and should be watched for in our region. Tomlinson (1969) has monographed the Acrothoracica on a worldwide basis.

Tomlinson, J. T. 1959. A burrowing barnacle of the genus *Trypetesa* (Order Acrothoracica). J. Wash. Acad. Sci., 43:373-81.

------. 1969. The burrowing barnacles (Cirripedia: Order Acrothoracica). Bull. 296, U. S. Nat. Mus. v + 162 pp.

ORDER RHIZOCEPHALA

The following species have either been reported from our area or may be expected to occur.

SUBORDER KENTROGONIDA

Angulosaccus tenuis Reinhard, 1944
on a hermit crab, *Parapagurus pilosimanus benedicti ("P. armatus")*, collected in very deep water off the coast of Washington

Clistosaccus paguri Lilljeborg, 1861
on hermit crabs: *Pagurus capillatus, P. dalli, Labidochirus splendescens* (Bering Sea, Alaska)

Peltogaster boschmae Reinhard, 1944
on the hermit crab *Discorsopagurus schmitti*

Peltogaster paguri Rathke, 1842
on hermit crabs: *Pagurus* spp.

Peltogasterella gracilis (Boschma, 1927) *(=P. socialis* Krüger, 1912; *=P. subterminalis* Reinhard, 1944)
on hermit crabs: *Pagurus* spp., *Elassochirus* spp., *Discorsopagurus schmitti*

Loxothylacus panopaei (Gissler, 1884)
on the crab *Lophopanopeus bellus*

Sylon hippolytes Krøyer, 1855
on shrimps of the genera *Hippolyte, Heptacarpus, Spirontocaris, Sclerocrangon, Pandalus*

SUBORDER AKENTROGONIDA

Thompsonia sp.
on the hermit crab *Discorsopagurus schmitti*

Reinhard, E. G. 1944. Rhizocephalan parasites of hermit crabs from the northwest Pacific. J. Wash. Acad. Sci., 34:49-58.

Reischman, P. G. 1959. Rhizocephala of the genus *Peltogaster* from the coast of the state of Washington and the Bering Sea. Proc. Nederl. Akad. Wetensch., C, 62:409-35.

SUBCLASS MALACOSTRACA

SUPERORDER PHYLLOCARIDA

ORDER LEPTOSTRACA

In our area, *Nebalia pugettensis* (Clark, 1932) is abundant in the lower intertidal and also subtidally. It prefers situations where algae and other organic detritus are decomposing. Clark originally described it as *Epinebalia pugettensis*, on the basis of the fact that the antenna of the male is sickle-shaped instead of straight as in species of *Nebalia*. However, the importance of this feature for segregation of genera is questioned.

Another species of *Nebalia*, in which the antenna of the male is straight, also occurs in this region; it is evidently undescribed. The females of the 2 species are almost indistinguishable.

Cannon, H. G. 1960. Leptostraca. *In* Bronn, H. G., Klassen und Ordnungen

des Tierreichs, 5, Abt. 1, Buch 4, Teil I. 81 pp.
Clark, A. E. 1932. *Nebaliella caboti* n. sp., with observations on other Nebaliacea. Trans. Royal Soc. Canada, Sect. V, ser. 3, 26:217-35.

SUPERORDER PERACARIDA

ORDER MYSIDACEA*

Banner (1948-1950, 1954a, 1954b) reported about 35 species of mysids from the northeastern Pacific region (Alaska to California). Some of these are strictly oceanic; others are found in plankton close to shore or in benthic situations, and a few are littoral. A mysid that is abundant on sandy beaches, in which it buries, is *Archaeomysis grebnitzkii* Czerniavsky, 1882, or a species close to it. (See choice 2b.)

Part III (1950) of Banner's monograph provides a comprehensive key to northeastern Pacific mysids. The works of Tattersall (1932, 1933, 1951), Tattersall & Tattersall (1951), and Ii (1964) will be helpful to anyone concerned with the systematics of these crustaceans, many of which have a wide distribution. The key below is designed to separate the species likely to be encountered in our region, especially in inland waters and in shore collecting.

1a Telson cleft 2
1b Telson not cleft 3
2a Antennal scale slender and tapering, setose all around (in eelgrass) *Mysis oculata* (Fabricius, 1780)
2b Antennal scale only about 3 times as long as wide, more or less truncate except for a tooth in the anterolateral corner, setose only on the anterior and mesial margins (in sand, intertidal) *Archaeomysis grebnitzkii* Czerniavsky, 1882
Ii (1964) indicated that mysids we refer to this species are sufficiently different to warrant description as a new species.
3a Antennal scale 4 to 5 times as long as wide, bearing a tooth in the anterolateral corner, and without setae on the lateral margin (in plankton) *Holmesiella anomala* Ortmann, 1908
3b Antennal scale elongated and tapering, at least 5 times as long as wide, without a tooth, and setose all around 4
4a Tip of antennal scale sharply pointed 5
4b Tip of antennal scale more or less rounded 7
5a Telson about twice as long as wide, approximately as long as the sixth abdominal segment; width of tip of telson at least one-tenth the width of the base (among *Ulva*, in bays) *Neomysis awatschensis* (Brandt, 1851)
5b Telson at least 2 and one-half times as long as wide, decidedly longer than the sixth abdominal segment; width of tip of telson less than one-tenth the width of the base 6
6a Telson widest a short distance distal to its base, distinctly truncate at the tip; marginal spines near distal end of telson longer than the distance between their bases, thereby appearing crowded (in plankton) *Neomysis integer* (Leach, 1815)
6b Telson widest at its base, truncate at the tip but not conspicuously so because of the narrowness of the distal portion; marginal spines near distal end of telson shorter than the distance between their bas-

*The key is a revision of one prepared by Laurel Kluge.

es, thereby appearing widely spaced (in plankton)
Neomysis rayii (Murdoch, 1885)

7a Apex of telson rounded 8

7b Apex of telson truncate (even if very narrow at the tip), not rounded 9

8a Eye not as long as the antennal scale (in plankton)
Acanthomysis pseudomacropsis (Tattersall, 1933)

8b Length of eye about equal to that of antennal scale (in plankton; also among *Ulva,* in bays) *Acanthomysis macropsis* (Tattersall, 1932)

9a Marginal spines of terminal quarter of telson conforming to the pattern observed in third quarter, with 1 to several smaller spines alternating with larger spines; tip of telson provided with 4 posteriorly-directed spines 10

9b Marginal spines of terminal quarter of telson abruptly deviating from the pattern observed in third quarter, consisting entirely of smaller spines of nearly equal length; tip of telson provided with 2 posteriorly-directed spines (in plankton; also in eelgrass)
Acanthomysis davisi Banner, 1948

10a Inner pair of spines at tip of telson longer than the outer pair, and generally decidedly longer than any of the marginal spines (in plankton; also in eelgrass and among *Ulva,* in bays)
Acanthomysis sculpta (Tattersall, 1933) and var. *nuda* Banner, 1948

10b Inner pair of spines at tip of telson shorter than the outer pair, and much shorter than the longest marginal spines (in plankton)
Acanthomysis nephrophthalma Banner, 1948

Banner, A. H. 1948. A taxonomic study of the Mysidacea and Euphausiacea (Crustacea) of the northeastern Pacific. Part I. Mysidacea, from Family Lophogastridae through Tribe Erythropini. Trans. Royal Canad. Inst., 26(1947):345-99.

------. 1948. A taxonomic study of the Mysidacea and Euphausiacea (Crustacea) of the northeastern Pacific. Part II. Mysidacea, from Tribe Mysini through Subfamily Mysidellinae. Trans. Royal Canad. Inst., 27:65-125.

------. 1950. A taxonomic study of the Mysidacea and Euphausiacea (Crustacea) of the northeastern Pacific. Part III. Euphausiacea. Trans. Royal Canad. Inst., 28(1949):1-63 [includes keys and bibliography for both Mysidacea and Euphausiacea].

------. 1954a. New records of Mysidacea and Euphausiacea from the northeastern Pacific and adjacent areas. Pacific Sci., 8:125-39.

------. 1954b. A supplement to W. M. Tattersall's review of the Mysidacea of the United States National Museum. Proc. U. S. Nat. Mus., 103:575-83.

Ii, N. 1964. Mysidae (Crustacea). *In* Fauna Japonica. Tokyo: Biogeographical Society of Japan, National Science Museum. x + 610 pp.

Tattersall, W. M. 1932. Contributions to a knowledge of the Mysidacea of California. Univ. Calif. Publ. Zool., 37:301-47.

------. 1933. Euphausiacea and Mysidacea from western Canada. Contrib. Canad. Biol. Fish., 8:183-205.

------. 1951. A review of the Mysidacea of the United States National Museum. Bull. 201, U. S. Nat. Mus. x + 292 pp.

Tattersall, W. M., & O. S. Tattersall. 1951. The British Mysidacea. London: Ray Society. viii + 460 pp.

ORDER CUMACEA

Josephine F. L. Hart
Victoria, British Columbia

The Cumacea of our area have been little studied. When Calman (1912) reported on the collections in the United States National Museum, there was not much material from Oregon, Washington, and British Columbia. Hart (1930) recorded 17 species from various localities around Vancouver Island. Zimmer (1936, 1943) reported 4 species from Puget Sound, and some that he described from other areas have since been found in the region covered by this manual. Wieser (1956, 1959) reported 3 species, and Lie (1968, 1969, 1971) recorded 19 species. Altogether, these papers account for 31 species of 11 genera in the area between 47° and 50° north latitude. Other species have been collected but not yet dealt with in publications.

Due to the present state of our knowledge, keys to the genera would seem to be the most useful. Many species are widely distributed, and for this reason the key provided includes some genera which have not yet been recorded here but which may be expected because they have been found either to the north or to the south of our area. For those wishing more detail, Jones (1963) has an excellent general account of the characteristics of the Order Cumacea; papers by Given (1961, 1964) and Lomakina (1958) should also be consulted.

Two keys are required because of sexual dimorphism. As the males mature, various structural differences develop that distinguish them from females and immature individuals. Males characteristically have large antennae; these eventually have articulated flagella as long as the body. Males may also have pleopods and larger, or additional, exopodites on the thoracic appendages. Females and immature individuals have rudimentary antennae, no pleopods, and fewer and less well developed exopodites. Males of the genera *Lamprops*, *Mesolamprops*, and *Hemilamprops* can be segragated on the basis of the absence of pleopods or the number of pleopods, but females and immature specimens show no such consistent differences.

Males

1a	With a distinct telson	2
1b	Without a telson	9
2a	Telson with 3 or more terminal spines	3
2b	Telson with 2 or no terminal spines	5
3a	Pleopods present	4
3b	Pleopods absent	*Lamprops*
4a	With 2 pairs of pleopods	*Mesolamprops*
4b	With 3 pairs of pleopods	*Hemilamprops*
5a	Telson with 2 terminal spines	6
5b	Telson without terminal spines	8
6a	Fourth thoracic segment, as viewed from above, longer than wide	*Diastylopsis*
6b	Fourth thoracic segment, as viewed from above, not longer than wide	7
7a	Telson short, bulbous, with not more than 1 pair of lateral spines	*Leptostylis*
7b	Telson long and tapered distally, usually with numerous lateral spines	*Diastylis*

8a Telson short, rounded, and not reaching beyond the anal valves *Colurostylis*

8b Telson long, reaching beyond the anal valves and tapering to an acute and upturned tip *Oxyurostylis*

9a Pleopods present 10

9b Pleopods absent 16

10a With 2 pairs of pleopods 11

10b With 3 or 5 pairs of pleopods 13

11a Carapace acute anteriorly *Leucon*

11b Carapace truncate anteriorly 12

12a Uropods with the exopodite shorter than the endopodite *Eudorella*

12b Uropods with the exopodite longer than the endopodite *Eudorellopsis*

13a With 3 pairs of pleopods *Leptocuma*

13b With 5 pairs of pleopods 14

14a Exopodite on pereopod 1 only *Cyclaspis*

14b Exopodites on the first 4 pereopods 15

15a Eye pigmented *Vaunthompsonia*

15b Eye not pigmented *Bathycuma*

16a Carapace bulbous and extending backwards over the free thoracic segments; anterolateral angle of the carapace usually rounded *Campylaspis*

16b Carapace not bulbous and not extending backwards over the free thoracic segments; anterolateral angle of the carapace acute *Cumella*

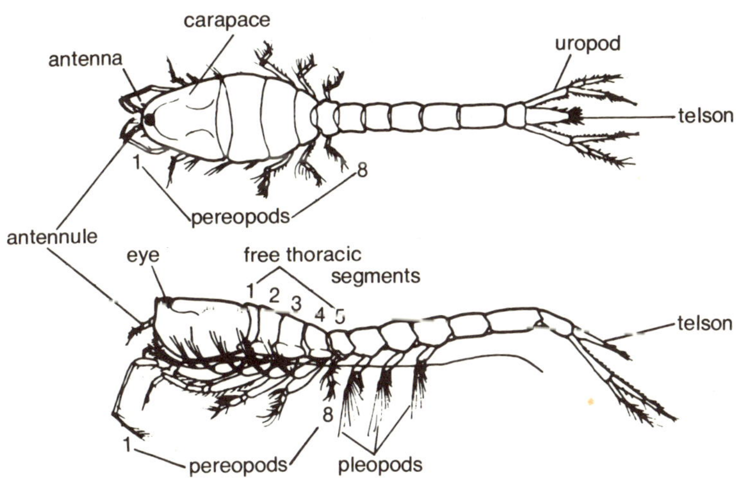

Females and Immature Males

1a With a distinct telson 2

1b Without a telson 7

2a Telson with 3 or more terminal spines *Lamprops, Hemilamprops, Mesolamprops*

2b Telson with 2 or no terminal spines 3

3a Telson with 2 terminal spines 4

3b Telson with no terminal spines 6

4a Fourth thoracic segment, as viewed from above, longer than wide *Diastylopsis*

4b Fourth thoracic segment, as viewed from above, not longer than wide 5

5a Telson short, somewhat bulbous, with no more than 1 pair of lateral spines *Leptostylis*

5b Telson long, tapered distally, with a number of lateral spines *Diastylis*

6a Telson short, not reaching beyond the anal valves *Colurostylis*

6b Telson long, reaching beyond the anal valves and tapering to an acute and upturned tip *Oxyurostylis*

7a With a double row of spines or spinules on the dorsal crest of the carapace 8

7b Without a double row of spines or spinules on the dorsal crest of the carapace 9

8a Eye pigmented *Vaunthompsonia*

8b Eye not pigmented *Bathycuma*

9a Exopodites on first pair of pereopods only *Cyclaspis*

9b Exopodites on at least the first 2 pairs of pereopods 10

10a Exopodites on the first and second pairs of pereopods 11

10b Exopodites on the first, second, and third pairs of pereopods 12

11a Carapace bulbous, extending backwards over the free thoracic segments; anterolateral angle of the carapace usually rounded *Campylaspis*

11b Carapace not bulbous and not extending backwards over the free thoracic segments; anterolateral angle of the carapace acute *Cumella*

12a Carapace acute anteriorly 13

12b Carapace truncate anteriorly 14

13a Eye present *Leptocuma*

13b Eye absent *Leucon*

14a Uropods with the exopodite shorter than the endopodite *Eudorella*

14b Uropods with the exopodite longer than the endopodite *Eudorellopsis*

Barnard, J. L., & R. R. Given. 1961. Morphology and ecology of some sublittoral cumacean Crustacea of southern California. Pacific Nat., 2: 153-65.

Calman, W. T. 1912. The Crustacea of the order Cumacea in the collections of the United States National Museum. Proc. U. S. Nat. Mus., 41:603-76.

Given, R. R. 1961. The cumacean fauna of the southern California continental shelf. No. 1. Family Leuconidae. Bull. So. Calif. Acad. Sci., 60: 130-46.

------. 1964. The cumacean fauna of the southern California continental shelf. No. 2. The new family Mesolampropidae. Crustaceana, 7:284-92.

Hart, J. F. L. 1930. Some Cumacea of the Vancouver Island region. Contrib. Canad. Biol. Fish., n. s., 6:25-40.

Jones, N. S. 1963. The marine fauna of New Zealand: Crustaceans of the order Cumacea. Bull. New Zealand Dept. Sci. Industr. Res., no. 152. 80 pp.

Lie, U. 1968. A quantitative study of benthic infauna in Puget Sound, Washington, U. S. A., in 1963-1964. Fisk Dir. Skr. (ser. Havunders.), 14:229-556.

------. 1969. Cumacea from Puget Sound and off the northwestern coast of Washington, with descriptions of 2 new species. Crustaceana, 17:19-30.

------. 1971. Additional Cumacea from Washington, U. S. A., with descrip-

tion of a new species. Crustaceana, 21:33-6.
Lomakina, N. B. 1958. [Cumacean crustaceans (Cumacea) of the seas of the U.S.S.R.] Opredeliteli po Faune SSSR, 66. 301 pp. (in Russian).
Sars, G. O. 1899-1900. Cumacea. *In* An Account of the Crustacea of Norway, with Short Descriptions and Figures of All the Species, vol. 3. 115 pp.
Stebbing, T. R. R. 1913. Cumacea (Sympoda). Das Tierreich, 39. 210 pp.
Wieser, W. 1956. Factors influencing the choice of substratum in *Cumella vulgaris* Hart (Crustacea, Cumacea). Limnol. Oceanogr., 1:274-85.
------. 1959. Free-living nematodes and other small invertebrates of Puget Sound beaches. Univ. Wash. Publ. Biol., 19. x + 179 pp.
Zimmer, C. 1936. California Crustacea of the order Cumacea. Proc. U. S. Nat. Mus., 83:423-39.
------. 1943. Cumaceen des Stillen Ozeans. Arch. Naturgesch., n. F., 12: 130-74.

ORDER TANAIDACEA

1a Pleon with 6 segments (including the pleotelson) 2
1b Pleon with 4 segments (including the pleotelson) *Pancolus californiensis* Richardson, 1905
2a With 3 pairs of pleopods; uropods uniramous *Anatanais normani* Richardson, 1905
2b With 5 pairs of pleopods; uropods biramous (but the exopodite may be very small) 3
3a Length of the first pereonite less than one-third the length of the fourth pereonite; movable and stationary claws of the chelae slender, about equal, and lacking teeth *Pseudotanais oculatus* Hansen, 1913
3b Length of the first pereonite at least half the length of the fourth pereonite; movable and stationary claws of the chelae either decidedly unequal or toothed 4
4a Endopodite of uropod with 4 articles; eyes present, pigmented; common intertidally on muddy substrates, especially where growths of *Ulva, Enteromorpha,* and related green algae are abundant *Leptochelia savignyi* (Krøyer, 1842)
4b Endopodite of uropod with 2 articles; eyes absent; dredged at moderate depths (10 to 20 m) *Leptognathia gracilis* (Krøyer, 1842)

Fee, A. R. 1927. The Isopoda of Departure Bay and vicinity, with descriptions of new species, variations, and colour notes. Contrib. Canad. Biol. Fish., n. s., 3:13-47.
Hatch, M. H. 1947. The Chelifera and Isopoda of Washington and adjacent regions. Univ. Wash. Publ. Biol., 10:155-274.
Lang, K. F. 1957. Tanaidacea from Canada and Alaska. Contr. Départ. Pêch. Québec, 52:1-54.
------. 1961. Further notes on *Pancolus californiensis* Richardson. Arkiv för Zoologi, ser. 2, 13:573-7.
Menzies, R. J. 1953. The apseudid Chelifera of the eastern tropical and north temperate Pacific Ocean. Bull. Mus. Comp. Zool., Harvard, 107: 442-96.
Richardson, H. 1905. A monograph of the isopods of North America. Bull. 54, U. S. Nat. Mus. lii + 727 pp.
------. 1905. Description of a new genus of Isopoda belonging to the family Tanaidae and of a new species of *Tanais*, both from Monterey Bay,

California. Proc. U. S. Nat. Mus., 28:367-70.

ORDER ISOPODA*

1a Adults parasitic on decapod crustacea; female generally much larger than the male, often asymmetrical and with reduced appendages; male symmetrical, and often living in close association with the female Suborder Epicaridea

1b Not parasitic on crustacea (usually free-living; some species are ectoparasites of fishes, and some are ectocommensal on crustacea, but these are not bizarrely modified as are the Epicaridea) 2

2a Uropods either lateral or ventral to the pleotelson 3

2b Uropods terminal 5

3a Uropods lateral, flattened, forming a caudal fan with the pleotelson 4

3b Uropods ventral, forming a sort of operculum covering the pleopods Suborder Valvifera

4a Body at least 7 times as long as wide, nearly cylindrical; exopodite of uropods arching over the pleotelson; pereopod 1 larger than the others and subchelate (pereopods 2 and 3 may also be subchelate) Suborder Anthuridea

4b Body usually not more than 5 times as long as wide; exopodite of uropods not arching over the pleotelson; pereopod 1 neither especially enlarged nor subchelate Suborder Flabellifera

5a Body slender, at least 7 times as long as wide; pleon (including the pleotelson) consisting of 3 nearly equal segments; pereopod 1 subchelate; small, generally under 2 mm long; living between sand grains Order Microcerberidea

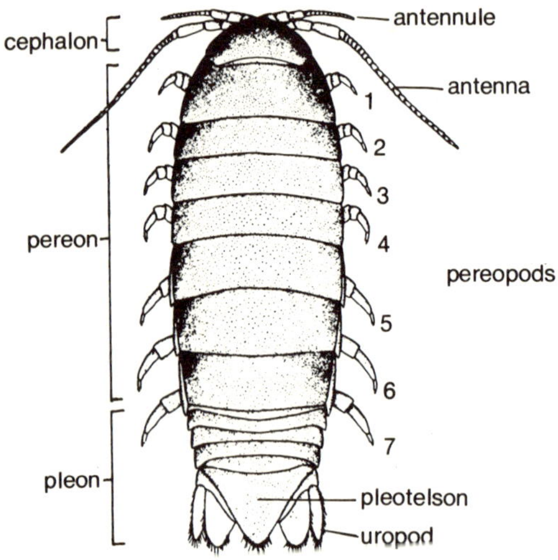

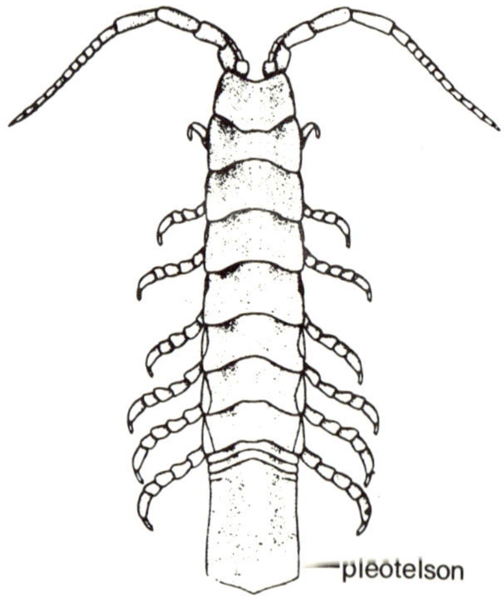

*The keys are based largely on information assembled by Jarl-Öve Stromberg and Robert Y. George.

5b Body usually robust, much less than 7 times as long as wide; pleon (including the pleotelson) not consisting of 3 nearly equal segments (if there are 3, the terminal one [pleotelson] is much larger than the other 2); pereopod 1 generally not subchelate (but sometimes subchelate); mostly more than 2 mm long (but some are smaller); in various habitats 6

6a Aquatic; pleon (including pleotelson) generally consisting of only 1 or 2 apparent segments (sometimes 3); both antennules and antennae well developed Suborder Asellota

6b Essentially terrestrial, although certain species are restricted to moist situations near the high tide line; pleon (including the pleotelson) generally consisting of 6 segments; antennules rudimentary, with only 2 or 3 articles Suborder Oniscoidea

SUBORDER EPICARIDEA

1a Female with recognizable pereopods, and some or all of the other appendages. Male distinct from the last larval stage (cryptoniscus) of the female 2

1b Female extremely modified, sac-like, generally without pereopods (when these are present, they are not jointed), and with most or all of the other appendages absent. Protandric hermaphrodites, the last larval stage (cryptoniscus) of the female serving as the male (hyperparasitic on the rhizocephalans *Peltogaster paguri* and *Peltogasterella gracilis*, which are parasitic on hermit crabs) *Liriopsis pygmaea* (Rathke, 1843)

2a Head of female with 2 dorsal hemispheres; pereopods of female rudimentary. Male lacking seventh pair of pereopods (parasitic in hepatic tissue of the crabs *Hemigrapsus oregonensis* and *H. nudus*) *Portunion conformis* Muscatine, 1956

2b Head of female not elaborated into 2 dorsal hemispheres; pereopods fairly well developed, though most of those on 1 side may be lacking. Male with 7 pairs of pereopods 3

3a Female with 1 side swollen and therefore longer than other side; only 1 pereopod (the first) present on longer side; pleon composed of 5 segments (parasitic on shrimps of the genera *Lebbeus*, *Spirontocaris*, *Eualus*, and *Heptacarpus*) *Hemiarthrus abdominalis* (Krøyer, 1840)

3b Both sides of female approximately equal; all pereopods present on both sides; pleon usually composed of 6 segments 4

4a Pleon of female with the lateral plates (pleural lamellae) elongated, fringed on 1 side by long, branched processes 5

4b Pleon of female with the lateral plates (pleural lamellae) poorly developed, without branched processes 6

5a Endopodite of pleopods much larger than exopodite, elongated and pointed; surface of endopodite roughened by irregularly transverse rugae; pleural plates of the last 3 pereon segments not developed as lamellae; distal portions of first incubatory lamellae forming a small lobe which is strongly curved inward (parasitic on hermit crabs) *Pseudione giardi* Calman, 1898

5b Endopodite of pleopods only a little larger than exopodite, triangular or ovate; surface of endopodite smooth; pleural plates of the last 3 pereon segments developed as lamellae; distal portions of first incubatory lamellae forming a large lobe which is directed straight backward (parasitic on the galatheid *Munida quadrispina*) *Pseudione galacanthae* Hansen, 1897

6a Pleopods present in female 7

6b Pleopods absent in female (parasitic on shrimps of the genera *Spirontocaris*, *Eualus*, and *Heptacarpus*) *Bopyroides hippolytes* (Krøyer, 1838)

7a Endopodites and exopodites of pleopods of adult female similar in size and shape, narrow and elongated. Male with segments of pleotelson distinct and with a pair of elongated pleopods on each of the first 5 segments; uropods present in male. First abdominal segment in both sexes bearing 2 dorsal papillae (parasitic on the anomuran *Upogebia pugettensis*) *Phyllodurus abdominalis* Stimpson, 1857

7b Endopodites and exopodites of pleopods of adult female dissimilar, the exopodite being narrow and elongated and the endopodite being oval and small. Male with segments of pleotelson fused, and without pleopods; uropods lacking. First abdominal segment in both sexes without dorsal papillae (parasitic on shrimps of the genera *Crangon*, *Argis*, and *Spirontocaris*) *Argeia pugettensis* Dana, 1853

SUBORDER VALVIFERA

1a Lateral margins of cephalon with deep incision; eyes decidedly dorsal; in brackish or essentially fresh water *Saduria entomon* (Linnaeus, 1767)

1b Lateral margins of cephalon without a deep incision; eyes essentially lateral; largely marine 2

2a Pleotelson consisting of a single segment, but showing lateral sutures of another partly coalesced segment; epimera of pereon coalesced with the segments; palp of maxilliped consisting of 3 articles 3

2b Pleon consisting of 3 segments (including the pleotelson) but showing lateral sutures of another partly coalesced segment; at least some of the epimera of the pereon distinct from the segments; palp of maxilliped consisting of 4 to 5 articles 7

3a Pleotelson becoming gradually narrowed posteriorly, but rounded at the apex, not notched or concave *Synidotea nodulosa* (Krøyer, 1848)

3b Apex of pleotelson notched or at least slightly concave 4

4a Width of body about three-tenths its length; width of pleotelson about three-fourths its length *Synidotea angulata* Benedict, 1897

4b Width of body more than one-third its length; width of pleotelson more than three-fourths its length and usually about as wide as long 5

5a Dorsal surface of each pereonite with a slight but distinct transverse ridge (on most of the pereonites this is near the posterior border), but without prominent tubercles *Synidotea bicuspida* (Owen, 1839)

5b Dorsal surface of pereonites with tubercles, at least on their lateral portions, but without a transverse ridge 6

6a Pereonites with 3 tubercles, 1 of which is on the midline; preocular horns of cephalon directed anterolaterally *Synidotea pettiboneae* Hatch, 1947

6b Pereonites with several tubercles, but all of them decidedly lateral to the midline; preocular horms of cephalon directed anteriorly *Synidotea ritteri* Richardson, 1904

7a Palp of maxilliped with 4 articles 7

7b Palp of maxilliped with 5 articles 11

8a Posterior margin of pleotelson with a median tooth or projection 9

8b Posterior margin of pleotelson concave *Idotea (Idotea) rufescens* Fee, 1926

9a Epimera of only the seventh pereonite extending to the posterior margin of the segment *Idotea (Idotea) fewkesi* Richardson, 1905

9b Epimera of both the sixth and seventh pereonites extending to the posterior margins of the segments 10

10a Posterolateral corners of pleotelson rounded, and the margins of the pleotelson between the corners and the tip of the apical projection slightly concave; length up to about 4 cm *Idotea (Idotea) ochotensis* Brandt, 1851

10b Posterolateral corners of pleotelson rather sharply angled, and the margins of the pleotelson between the corners and the tip of the apical projection somewhat sinuous; length up to about 2.5 cm *Idotea (Idotea) urotoma* Stimpson, 1864

11a Posterior margin of pleotelson convex, often with a median apical projection 12

11b Posterior margin of pleotelson concave *Idotea (Pentidotea) resecata* (Stimpson, 1857)

12a Outline of eyes somewhat pear-shaped; posterolateral margin of epimeral plate of seventh pereonite acute 13

12b Outline of eyes circular, oval, elongated, or kidney-shaped; posterolateral margin on epimeral plate of seventh pereonite rounded, not acute *Idotea (Pentidotea) schmitti* Menzies, 1951

13a First pleonite shorter laterally than medially; outline of eyes kidney-shaped *Idotea (Pentidotea) wosnesenskii* (Brandt, 1851)

13b First pleonite at least as long laterally as medially; outline of eyes circular, oval, or kidney-shaped 14

14a Outline of eyes oval or circular; maxilliped with 1 coupling hook; length up to about 2.5 cm *Idotea (Pentidotea) aculeata* (Stafford, 1913)

14b Outline of eyes elongated; maxilliped with 2 coupling hooks; length up to about 5.5 cm *Idotea (Pentidotea) stenops* (Benedict, 1898)

SUBORDER ANTHURIDEA

A paper by Menzies (1951) includes descriptions of 3 species from California. Any of these, as well as others, may be expected in our region. These small isopods, generally under 3 mm long, should be looked for in deposits of sand and other gritty material accumulating on or in holdfasts of kelps, and in similar situations.

SUBORDER FLABELLIFERA

1a Pleon (including the pleotelson) composed of 5 or 6 distinct segments 2

1b Pleon appearing to consist of only 2 segments, due to fusion of the first 5 10

2a Uropods with both rami well developed, usually lamellar; not capable of rolling up into a ball; not boring in wood 3

2b Uropods with the exopodite rudimentary and more or less clawlike; capable of rolling up into a ball; boring in wood *Limnoria lignorum* Rathke, 1799

3a Antennules and antennae with a well-defined peduncle and flagellum; pereopods 1 to 3, or none of the pereopods, prehensile 4

3b Antennules and antennae reduced, without any appreciable distinction between the peduncle and flagellum; all pereopods prehensile 16

4a Palp of maxilliped free; margins of last 2 articles of palp more or less setose, never armed with hooks; free-living 5
4b Palp of maxilliped embracing the cone formed by the mouth parts; apex of palp armed with hooks, never setose; ectoparasitic on fishes, but may also be found swimming free 6
5a Cephalon rounded in front; posterior margin of pleotelson rather smoothly rounded *Cirolana harfordi* Lockington, 1877
5b Cephalon with median spatuliform process; posterior margin of pleotelson forming an obtuse angle medially *Cirolana kincaidi* Hatch, 1947
6a Flagellum of antennule with numerous articles; propodus of prehensile pereopods not expanded *Aega symmetrica* Richardson, 1905
6b Flagellum of antennule with only 4 to 6 articles; propodus of prehensile pereopods expanded along the inner margin 7
7a Cephalon without lateral lobes in front of eyes; inner angle of peduncle of uropod usually as long as exopodite 8
7b Cephalon with lateral lobes in front of eyes, especially towards their outer angles; inner angle of peduncle of uropod only a little more than half as long as exopodite (ectoparasitic on halibut and *Raja binoculata)* *Rocinela angustata* Richardson, 1898
8a Propodus of prehensile pereopods arcuately lobed and spinulose along the inner margin; merus of prehensile pereopods spinulose 9
8b Propodus of prehensile pereopods with a distally curved process set with 6 blunt teeth meeting squarely and without interval; merus of prehensile pereopods with about 5 teeth (ectoparasitic on *Raja binoculata, Hippoglossus* sp., and rock cod) *Rocinela propodialis* Richardson, 1905
9a Median lobe of cephalon simple (ectoparasitic on various fishes, including *Hydrolagus colliei,* species of *Raja,* cod, rockfish, and halibut) *Rocinela belliceps* (Stimpson, 1864)
9b Median lobe of cephalon trifid (specific fish hosts not recorded) *Rocinela tridens* Hatch, 1947
10a Not capable of rolling up into a ball; first pair of pereopods subchelate in both sexes, second pair subchelate in the male *Tecticeps pugettensis* Hatch, 1947
10b Capable of rolling up into a ball; none of the pereopods subchelate in either sex 11
11a Pleotelson either trilobed or deeply notched at the apex 12
11b Pleotelson neither trilobed nor deeply notched at the apex 14
12a Pleotelson trilobed at the apex *Cymodoce japonica* Richardson, 1906
12b Pleotelson deeply notched at the apex 13
13a Dorsal surface of pleotelson with rows of tubercles *Dynamenella sheareri* (Hatch, 1947)
13b Dorsal surface of pleotelson smooth *Dynamenella glabra* (Richardson, 1897)
14a Three pleonites reaching the lateral margin; no respiratory folds on pleopods 4 and 5 *Gnorimosphaeroma oregonense* (Dana, 1855)
G. luteum Menzies, 1954 has only 2 pleonites that reach the margin.
14b Only one pleonite reaching the lateral margin; respiratory folds present on pleopods 4 and 5 15
15a Pleotelson triangular, its length greater than one-third the total body length; dorsal surface with tubercles *Exosphaeroma amplicauda* (Stimpson, 1857)
15b Pleotelson with broadly truncate posterior border, its length less than one-third the total body length; dorsal surface smooth

Exosphaeroma media George & Strömberg, 1968

16a Distance between the eyes equal to about 3 times their diameter; second antenna with 10 to 11 articles; anterior margin of cephalon broadly rounded or nearly truncate; pleotelson (of adult female) almost twice as broad as long *Lironeca vulgaris* Stimpson, 1857

16b Distance between eyes equal to about 1 and one-half times their diameter; second antenna with 8 or 9 articles; anterior margin of cephalon with a blunt projection; pleotelson (of adult female) about as broad as long *Lironeca californica* Schiödte & Meinert, 1883-84

SUBORDER MICROCERBERIDEA

The members of this group are small and occur typically in substrates of sand and fine gravel. Only 1 species, *Microcerberus abbotti* Lang, 1961, has been described from the Pacific coast. Originally found in central California, it has since been reported from southern California. A subspecies found on San Juan Island, Washington, was named *M. abbotti juani* by Coineau and Delamare Deboutteville (1967).

SUBORDER ASELLOTA

1a Uropods minute, without peduncle, but almost always with endopodite and exopodite 9

1b Uropods generally either with a peduncle or absent (if the uropods consist of a single article, the body as a whole is either narrow and with nearly parallel margins, or there are posterolateral processes on the pleotelson) 2

2a Articles 1 to 5 of palp of maxilliped all narrow and of about equal width, less than half as wide as the endite *Jaeropsis setosa* George & Strömberg, 1968

2b Articles 1 to 3 of palp of maxilliped conspicuously wider than articles 4 to 5, at least as wide as the endite 3

3a Inferior margin of propodus of first pereopod dentate for about one-third of its length 4

3b Inferior margin of propodus of first pereopod lacking dentations 5

4a Rostrum of cephalon elongated and pointed; associated with the asteroid *Solaster stimpsoni* *Janiralata solasteri* (Hatch, 1947)

4b Rostrum of cephalon short, triangular; apparently free-living *Janiralata occidentalis* (Walker, 1898)

5a Pleotelson with 3 spinelike serrations on each side *Ianiropsis tridens* Menzies, 1952

5b Pleotelson with more than 3 spinelike serrations on each side 6

6a Pleotelson as long as, or longer than, wide 7

6b Pleotelson much wider than long *Ianiropsis analoga* Menzies, 1952

7a Length of uropods exceeding half the length of the pleotelson *Ianiropsis kincaidi* Richardson, 1904

7b Length of uropods not exceeding half the length of the pleotelson *Ianiropsis magnocula* Menzies, 1952

8a Cephalon wide, not enclosed by lateral portions of the first pereonite; eyes well developed, with numerous ommatidia 9

8b Cephalon small and narrow, enclosed by lateral portions of the first pereonite; eyes usually absent (but if present, separate ommatidia are not recognizable) 11

9a Uropods small and leaflike; pleotelson laterally serrated

Munna (Uromunna) ubiquita Menzies, 1952

9b Uropods not leaflike (rounded in transverse section); pleotelson without lateral serrations 10

10a Uropods with at least 1 large, medially recurved apical spine; maxilliped with 3 coupling hooks; ventral surface of operculum of female with 2 longitudinal rows of setae *Munna (Neomunna) chromatocephala* Menzies, 1952

10b Uropods without recurved apical spines; maxilliped with 2 coupling hooks; ventral surface of operculum of female with numerous setae, these not in 2 rows *Munna (Munna) fernaldi* George & Strömberg, 1968

11a Eyes absent; epimera pointed and clearly visible in dorsal view *Pleurogonium rubicundum* Sars, 1863

11b Eyes present; epimera rounded and hardly visible in dorsal view *Munnogonium waldronense* George & Strömberg, 1968

SUBORDER ONISCOIDEA

1a Flagellum of antenna with more than 10 articles (there may be as few as 6 articles in immature specimens); male copulatory apparatus involving only the endopodites of the second pleopods, and these not fused medially; penial processes, if present, paired; up to about 2.5 cm long; found in rock crevices just above high tide line *Ligia pallasii* Brandt, 1833

1b Flagellum of antenna with no more than 5 articles; male copulatory apparatus frequently involving the endopodites of both first and second pleopods, those of the first pleopods fused with a median appendage; penial processes, if present, unpaired; less than 2 cm long; not typically found in rock crevices (generally on uppermost portions of sandy beaches or truly terrestrial) 2

2a Tracheal system absent; flagellum of antenna with 3 or 4 articles; under debris on sandy beaches, near uppermost limits of intertidal zone 4

2b Tracheal system present (evident in exopodites of first 2 pleopods); flagellum of antenna with 2 articles; truly terrestrial, though occasionally found close to the uppermost limits of the intertidal zone 5

3a Not capable of rolling up into a ball; antennae extending posteriorly beyond second pereonite; pleotelson pointed *Porcellio scaber* Latreille, 1804

3b Capable of rolling up into a ball; antennae not extending posteriorly beyond second pereonite; pleotelson blunt *Armadillidium vulgare* (Latreille, 1804)

4a Flagellum of antenna with 3 articles 5

4b Flagellum of antenna with 4 articles 6

5a Pleonites becoming gradually narrower than the pereonites; cephalon with large and tuberculate median and anterolateral lobes; antennae not extending beyond first pereonite *Alloniscus perconvexus* Dana, 1854

5b Pleonites becoming abruptly narrower than the pereonites; cephalon rounded in front, lateral lobes very small if present; antennae long, extending well beyond second pereonite *Philoscia (Littorophiloscia) richardsonae* Holmes & Gay, 1909

6a Anterior margin of cephalon very deeply 3-lobed, the median lobe forming a rostrum; basopodite of uropods expanded, forming a large lobe lateral to the endopodite and nearly reaching the distal tip of the latter *Armadilloniscus tuberculatus* Holmes & Gay, 1909

6b Anterior margin of cephalon 2-lobed, without a rostrumlike median lobe; basopodite of uropod not expanded, the endopodite inserted upon it in a posterolateral position *Detonella papillicornis* Richardson, 1904

Coineau, N., & C. Delamare Deboutteville. 1967. Étude des microcerbérides (Crustacés, Isopoda) de la côte Pacifique des États-Unis. Ire partie: Systématique. Bull. Mus. Nat. Hist. Nat., sér. 2, 39:955-64.

George, R. Y., & J.-O. Strömberg. 1968. Some new species and new records of marine isopods from San Juan Archipelago, Washington, U. S. A. Crustaceana, 14:225-54.

Hatch, M. H. 1947. The Chelifera and Isopoda of Washington and adjacent regions. Univ. Wash. Publ. Biol., 10:155-274.

Lang, K. 1961. Contributions to the knowledge of the genus *Microcerberus* Karaman (Crustacea: Isopoda) with a description of a new species from the central Californian coast. Arkiv för Zoologi, ser. 2, 13:493-510.

Menzies, R. J. 1950. The taxonomy, ecology, and distribution of northern California isopods of the genus *Idothea* with the description of a new species. Wasmann J. Biol., 8:155-95.

------. 1951. New marine isopods, chiefly from northern California, with notes on related forms. Proc. U. S. Nat. Mus., 101:105-56.

------. 1952. Some marine asellote isopods from northern California, with descriptions of nine new species. Proc. U. S. Nat. Mus., 102: 117-59.

Menzies, R. J., & M. A. Miller. 1972. Systematics and zoogeography of the genus *Synidotea* (Crustacea: Isopoda). Smithsonian Contr. Zool., no. 102. 33 pp.

Muscatine, L. 1956. A new entoniscid (Crustacea: Isopoda) from the Pacific coast. Jour. Wash. Acad. Sci., 46:122-6.

Richardson, H. 1905. A monograph of the isopods of North America. Bull. 54, U. S. Nat. Mus. liii + 727 pp.

Schultz, G. A. 1969. How to know the Marine Isopod Crustaceans. Dubuque, Iowa: W. C. Brown Co. vii + 359 pp.

Van Name, W. G. 1936. American land and fresh water isopod Crustacea. Bull. Amer. Mus. Nat. Hist., 71:1-535.

ORDER AMPHIPODA

1a Pereon consisting of 7 apparent segments, all provided with well developed appendages; abdomen not vestigial; body neither especially slender nor resembling that of a praying mantis 2

1b Pereon consisting of 6 apparent segments, some of which may have only vestigial appendages; abdomen vestigial; body (except for species found on whales) usually slender, and often resembling a praying mantis Suborder Caprellidea

2a Eyes generally large, occupying most of the head; coxae of pereopods small and often fused with the body; usually planktonic, or associated with jellyfishes or occupying the tunic of dead salps Suborder Hyperiidea

2b Eyes usually present and conspicuous, but not so large as to occupy most of the head; coxae of pereopods well developed, and usually expanded into platelike structures; mostly benthic, but a few species are planktonic Suborder Gammaridea

SUBORDER HYPERIIDEA

These are generally either independently planktonic, or associated with jellyfishes. One remarkable genus in our area, *Phronima,* lives in empty tests of salps, and is sometimes collected in plankton tows or alongside floating docks. The hyperiids will not be treated in this manual, but the following recent references will lead one to the available literature pertinent to our fauna.

Bowman, T. E. 1973. Pelagic amphipods of the genus *Hyperia* and closely related genera (Hyperiidea: Hyperiidae). Smithsonian Contr. Zool., no. 136. 76 pp.

Bowman, T. E., & H.-E. Gruber. 1973. The families and genera of Hyperiidea (Crustacea: Amphipoda). Smithsonian Contr. Zool., no. 146. iv + 64 pp.

Shih, C.-T. 1969. The systematics and biology of the family Phronimidae (Crustacea: Amphipoda). Dana Reports, 74:1-100.

SUBORDER GAMMARIDEA

Although the gammaridean amphipods are an important part of the biomass in many marine and freshwater habitats, their taxonomy is so difficult that only a few zoologists seem to have the right combination of talent and determination to work on systematics of the group. Our fauna has been very inadequately studied. Thorsteinson (1941) reported or described a number of species occurring in this region. Barnard (1954) made a survey of gammarideans on the coast of Oregon, and Mills (1961, 1962) published studies dealing with 2 families represented on the Pacific coast of Canada and in a few localities in Washington. However, these seem to be the only recent papers of a reasonably comprehensive nature for this area.

The situation in California is much better. Barnard (1969b) prepared a valuable summary of the gammaridean amphipods of rocky intertidal habitats of central and southern California, and his section on intertidal species in *Light's Manual* (1974) is perhaps as practical an approach as is possible to a large and complex fauna. Barnard has also published papers on various families represented in California.

The review by Barnard (1969a) of the families and genera of gammaridean amphipods, on a worldwide basis, is a monumental achievement. This work is well illustrated and has keys to the genera, which makes it extremely useful in our region. A few taxa have been treated differently in *Light's Manual,* but as Barnard himself has pointed out, the taxonomy of the Amphipoda cannot very well be stable: "No specialist is nearly satisfied with current classification at familial levels. Many genera are extremely confused, not only for the lack of clear description of their type-species but for the obviously erroneous inclusion of many of their congeners."

Some other comprehensive works cited in the bibliography may be helpful. Guri͡anova (1951, 1962) dealt with many taxa represented in our area. The classical accounts of Sars (1895) and Stebbing (1906) remain important because of their descriptive illustrations of many type-species.

The key provided here is by no means definitive. It is, in fact, barely a provisional one. It is offered with the hope that it will help a nonspecialist recognize the families to which most gammaridean amphipods of our area belong, and thus to find the tracks leading to determination of genera and species. As our fauna becomes better understood, and as improvement and enlargement of the key is undertaken, we should in a few years be able to develop a treatment of amphipods comparable to that in *Light's Manual*. The latter, by the way, will be very useful in the Pacific Northwest because so many of the species found in California also occur in our area.

The key takes one to the family, and some of the genera likely to be found in our area are listed. In the case of families represented by only 1 known species (or only 1 common species, at least), this species is indicated. The characteristics used in the key to families are not necessarily characteristic of all genera of each of these families. Thus, unfortunately, a genus with traits unlike those of other genera in the same family may not fit the criteria for identification of that family. An effort, however, has been made to make most genera expected in this region go through the key without complications.

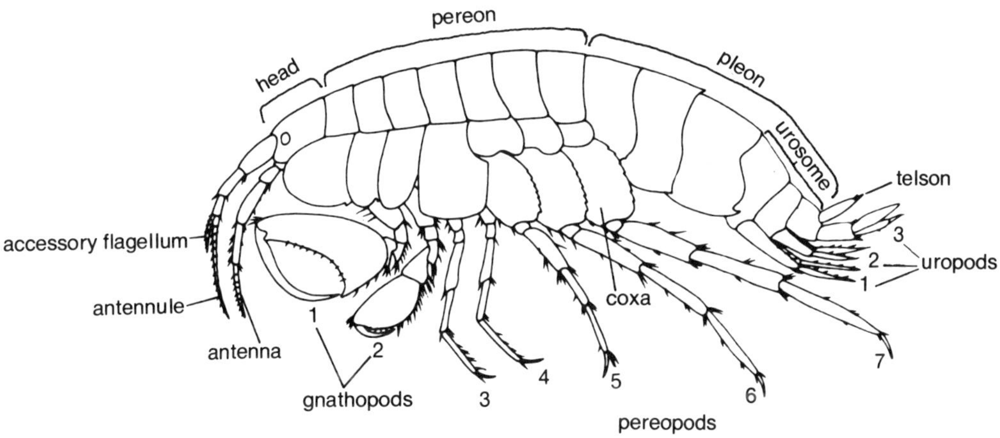

1a Dorsal and lateral surfaces of body with conspicuous spinelike processes; most coxae (at least coxae 1 to 4) ending in 1 or more sharp points 2
1b Dorsal and lateral surfaces without conspicuous spinelike processes; coxae usually not ending in sharp points 3
2a Mouthparts grouped into a cone-shaped bundle projecting from the ventral surface of the head; mandible with a poorly developed molar, or none Acanthonotozomatidae
2b Mouthparts not grouped into a cone-shaped bundle projecting from the ventral surface of the head; mandible with a well developed molar Paramphithoidae
3a Article 3 of gnathopod 2 elongate (generally at least 3 times as long as wide) Lysianassidae
Anonyx, *Aristias*, *Hippomedon*, *Koroga* (planktonic), *Orchomene*, *Tryphosella*
3b Article 3 of gnathopod 2 short 4
4a Eyes united medially 5
4b Eyes not united medially 6

5a Telson not cleft Oedocerotidae
Monoculodes, Synchelidium, Westwoodilla

See Mills (1962).

5b Telson cleft Synopiidae
Tiron

Tiron has a small secondary eye on each side of the head.

6a Eyes unusual, consisting of some internal cephalic masses (red pigment usually evident) and 4 corneal lenses (2 on each side of the head) Ampeliscidae
Ampelisca

6b Eyes typical, consisting of a single mass of compacted ommatidia on each side of the head 7

7a Telson somewhat swollen, not cleft; when the telson is manipulated, it generally changes shape, rather than moving as a unit with respect to its articulation with the urosome 8

7b Telson generally thin and somewhat flattened, cleft or uncleft; when the telson is manipulated, it generally moves as a unit with respect to its articulation with the urosome 13

8a Uropod 3 hidden beneath the telson (thus appearing to be absent) 9

8b Uropod 3 readily visible 10

9a Body almost cylindrical (resembling a tanaidacean); abdomen not bent under the thorax; boring in kelp Eophliantidae
Wandelia

9b Body somewhat flattened dorsoventrally; abdomen bent under the thorax; not boring in kelp Podoceridae
Dulichia, Podocerus

10a Outer ramus (or the only ramus) of uropod 3 with an apical hook or with 2 hooked spines 11

10b Outer ramus of uropod 3 without an apical hook or hooked spines 12

11a Outer ramus of uropod 3 stout, with 2 hooked spines; inner ramus flat, and with bristles near its apex Amphithoidae
Amphithoe

11b Outer ramus of uropod 3 slender, with an apical hook and also some small teeth; inner ramus slender and without bristles Ischyroceridae
Ischyrocerus, Jassa

12a Uropods 1 and 2 similar to one another in form, although uropod 1 is generally much larger, and none of the uropods have a leaflike peduncle; not boring in wood (however, some members of this group may be found in burrows made by other animals) Corophiidae
(Aoridae, Corophiidae, Isaeidae, Photidae)
Aoroides, Corophium, Ericthonius, Eurystheus, Gammaropsis, Microprotopus, Photis, Podoceropsis, Protomedia

12b All 3 uropods strikingly dissimilar, and the peduncle of uropod 2 distinctive in being leaflike and partially fringed by bristles; boring in wood Cheluridae
Chelura terebrans Philippi, 1839

13a Area of coxa 1 at least half that of coxa 2 14

13b Area of coxa 1 not more than about three-eighths that of coxa 2 15

14a Uropod 3 with 2 rami Amphilochidae
Amphilochus

14b Uropod 3 with a single ramus (thus it appears to consist of 3 articles in a row) Stenothoidae
Mesometopa, Metopa, Stenothoe

15a Segments 5 and 6 of pleon (urosome segments 2 and 3) fused 16
15b Segments 5 and 6 of pleon distinct and articulating freely 17
16a Mandible with a palp; not embedded in the tunic of compound ascidians
Atylidae
Atylus

See Mills (1961).

16b Mandible without a palp; our species embedded in the tunic of compound ascidians
Dexaminidae
Polycheria osborni Calman, 1898

17a Mandible without a palp; inner ramus of uropod 3 either absent or vestigial, the other ramus not longer than the peduncle
Superfamily Talitroidea

Dogielinotidae: *Dogielinotus*
Hyalidae: *Allorchestes, Hyale, Parallorchestes*
Najnidae: *Najna*
Talitridae: *Orchestia, Orchestoidea*

These families are difficult to separate in a key. See Barnard (1969a; also 1974, in *Light's Manual).*

17b Mandible with a palp; inner ramus of uropod 3 generally distinct, even if short (if the inner ramus is very small, then the outer ramus is longer than the peduncle) 18
18a Carpus (article 5) of gnathopod 1 extended into a long process that functions as a fixed finger against which the dactyl (article 7) closes
Leucothoidae
Leucothoë

18b Carpus of gnathopod 1 not extended into a long process that functions as a fixed finger against which the dactyl closes 19
19a Accessory flagellum of antennule either absent or consisting of a single article 20
19b Accessory flagellum of antennule consisting of at least 2 articles 21
20a Telson entire
Pleustidae
Neopleustes, Parapleustes, Pleustes

20b Telson cleft
Eusiridae
Pontogeneia

21a Gnathopods 1 and 2 both large and of similar size
Liljeborgiidae
Liljeborgia, Listriella

21b Gnathopods 1 and 2 either small or of very different sizes 22
22a Pereopod 7 shorter than pereopod 6, and of different form
Phoxocephalidae
Harpinia, Paraphoxus

22b Pereopod 7 long, and of the same form as pereopod 6 23
23a Pereopods 3 to 7 lacking dactyls; appendages with fringes of long bristles, thus appearing "furry"; burrowing in sand
Haustoriidae
Eohaustorius

E. washingtonianus (Thorsteinson, 1941) is common in our area, but other species should be expected.

23b Pereopods 3 to 7 with dactyls; appendages not usually fringed with such long bristles as to appear "furry"; not usually burrowing in sand (although some species do)
Gammaridae
Anisogammarus (brackish water), *Ceradocus, Elasmopus, Melita, Maera*

Barnard, J. L. 1954. Marine Amphipoda of Oregon. Oregon State Monogr.,

Zool., no. 8. 103 pp.
------. 1958. Index to the families, genera, and species of the gammaridean Amphipoda (Crustacea). Occ. Pap. Allan Hancock Found., no. 19. 145 pp.
------. 1969a. The families and genera of marine gammaridean Amphipoda (Crustacea). Bull. 271, U. S. Nat. Mus. vi + 535 pp.
------. 1969b. Gammaridean Amphipoda of the rocky intertidal of California: Monterey Bay to La Jolla. Bull. 258, U. S. Nat. Mus. x + 230 pp.
------. 1971. Gammaridean Amphipoda from a deep-sea transect off Oregon. Smithsonian Contr. Zool., no. 61. v + 86 pp.
Gur͡ianova, E. F. 1951. [Amphipods of the seas of the U. S. S. R. and adjacent waters (Amphipoda: Gammaridea).] Opredeliteli po Faune SSSR, 41. 1029 pp. (in Russian)
------. 1962. [Amphipods of the northern part of the Pacific Ocean, part 1.] Opredeliteli po Faune SSSR, 74. 440 pp. (in Russian)
Mills, E. L. 1961. Amphipod crustaceans of the Pacific coast of Canada, 1: Family Atylidae. Nat. Mus. Canada Bull. 172:13-33.
------. 1962. Amphipod crustaceans of the Pacific coast of Canada, 2: Family Oedocerotidae. Nat. Hist. Pap. Nat. Mus. Canada, 15:1-21.
Sars, G. O. 1895. Amphipoda. *In* An Account of the Crustacea of Norway with short descriptions and figures of all the species, vol. 1. vii + 711 pp.
Stebbing, T. R. R. 1906. Amphipoda, 1: Gammaridea. Das Tierreich, pt. 21. 806 pp.
Thorsteinson, E. D. 1941. New or noteworthy amphipods from the north Pacific coast. Univ. Wash. Publ. Oceanogr., 4:50-96.

SUBORDER CAPRELLIDEA

The key does not cover the genera and species found on whales (family Cyamidae). For these, consult the paper of Leung (1967), and also the account of Caprellidea in *Light's Manual*.

1a With gills on pereonites 2, 3, and 4; mandible without a molar process 2
1b With gills on pereonites 3 and 4 only; mandible with a molar process 3
2a Pereopods 3 and 4 with 1 article, pereopod 5 with 6 articles; abdomen with 5 segments *Cercops compactus* Laubitz, 1970
2b Pereopods 3 and 5 with 3 articles, pereopod 4 with 1 article; abdomen small and without obvious segmentation *Perotripus brevis* (La Follette, 1915)
3a With vestigial pereopods on pereonites 3 and 4; mandible with a palp 4
3b Without pereopods on pereonites 3 and 4; mandible without a palp 7
4a Pereopods 3 and 4 consisting of 2 articles; antennae without long setae 5
4b Pereopods 3 and 4 consisting of only 1 article; antennae with long setae on the larger articles 6
5a Head with a dorsal median spine (in the male) or knob (in the female); pereopod 5 consisting of 6 articles; gnathopod 2 of male attached to the anterior half of pereonite 2 *Deutella californica* Mayer, 1890
5b Head without a dorsal median spine or knob; pereopod 5 consisting of 3 articles; gnathopod 2 of male attached to the posterior half of pereo-

nite 2 *Mayerella banksia* Laubitz, 1970

6a Penultimate article of antenna about 4 times as long as wide, with setae not appreciably longer than its width; short spines on pereonites directed more nearly anteriorly than laterally *Tritella laevis* Mayer, 1903

6b Penultimate article of antenna about 6 times as long as wide, with some setae about twice as long as its width; short spines on pereonites directed more nearly laterally than anteriorly *Tritella pilimana* Mayer, 1890

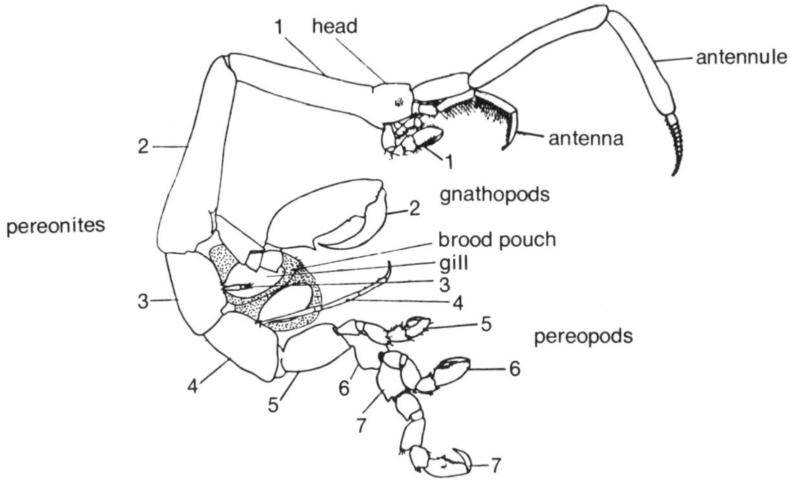

7a With a ventral spine between the bases of gnathopods 2 8

7b Without a ventral spine between the bases of gnathopods 2 11

8a With a prominent anteriorly directed spine on the dorsal surface of the head; propodus of gnathopod 2 of male about 4 times as long as wide *Caprella californica* Stimpson, 1857

8b Without a spine on the dorsal surface of the head; propodus of gnathopod 2 of male only about twice as long as wide 9

9a Pereonite 5 with a small but distinct lateral projection on both sides of its anterior half (there are also projections on the posterior half); with a conspicuous anterolaterally directed spine lateral to the base of gnathopod 2 *Caprella equilibra* Say, 1818

9b Pereonite 5 without a projection on both sides of its anterior half; if a spine is present lateral to the base of gnathopod 2, it is small and not directed anterolaterally 10

10a Dactyl of gnathopod 2 decidedly setose; without a spine lateral to the base of gnathopod 2 *Caprella pilidigita* Laubitz, 1970

10b Dactyl of gnathopod 2 not setose; with a small spine lateral to the base of gnathopod 2 *Caprella mendax* Mayer, 1903

11a Head without any dorsal spines, knobs, or tubercles 12

11b Head with one or a pair of dorsal spines, knobs, or tubercles 17

12a Propodus of pereopods 5 to 7 slender, its inner edge nearly straight and with a tooth near the middle; dactyl of gnathopod 2 of male decidedly setose; basis of gnathopod 2 of male much longer than the propodus *Caprella gracilior* Mayer, 1903

12b Propodus of pereopods 5 to 7 rather stout, much of its inner edge at least slightly concave (except perhaps in *C. striata*) and with a tooth proximal to its middle; dactyl of gnathopod 2 of male not setose; basis of gnathopod 2 of male shorter than the propodus 13

13a Propodus of gnathopod 2 with 2 small spines at the base of the large grasping spine; pereonite 1 of male about 3 times as long as the head *Caprella irregularis* Mayer, 1890

13b Propodus of gnathopod 2 with only 1 small spine at the base of the large grasping spine; pereonite 1 of male not more than twice as long as the head 14

14a Pereonites (at least the more posterior ones) with dorsal spines or tubercles; antennule more than half the length of the body in the female, more than two-thirds the length of the body in the male 15

14b Pereonites without dorsal spines or tubercles; antennule less than half the length of the body 16

15a Flagellum of antennule longer than the peduncle; antenna of male about as long as the peduncle of the antennule; subtidal *Caprella striata* Mayer, 1903

15b Flagellum of antennule shorter than the peduncle; antenna of male shorter than the combined length of the first 2 articles; has been taken intertidally *Caprella alaskana* Mayer, 1903

16a Outline of gills nearly circular; gnathopod 2 of male setose; gnathopod 2 of female attached near the middle of pereonite 2 (not likely to be found in our area) *Caprella drepanochir* Mayer, 1890

16b Outline of gills elongated; gnathopod 2 of male not setose; gnathopod 2 of female attached near the anterior end of pereonite 2 *Caprella leviuscula* Mayer, 1903

17a Head with a single dorsal spine or knoblike projection 19

17b Head with a pair of dorsal spines or tubercles 29

18a Dorsal spine on head an anteriorly-directed triangular projection 19

18b Dorsal projection on head a knob that is directed upward 21

19a Tubercles on dorsal surface small or absent (if present at all, lacking on pereonite 1); gnathopod 2 of male attached near the anterior end of pereonite 2 *Caprella angusta* Mayer, 1903

19b Tubercles on dorsal surface of body distinct, present on all pereonites; gnathopod 2 of male attached near the middle of pereonite 2 20

20a Tubercles on body large; propodus of gnathopod 2 shorter than pereonite 2 *Caprella verrucosa* Boeck, 1872

20b Tubercles on body small; propodus of gnathopod 2 of male as long as pereonite 2 *Caprella incisa* Mayer, 1903

21a Head and pereon covered with large and small tubercles; antenna with some long setae (more than twice as long as the articles from which they originate); gnathopod 2 and much of the body of the male setose *Caprella pustulata* Laubitz, 1970

21b Head and pereon with only a few tubercles; antenna without especially long setae (none more than twice as long as the width of the articles from which they originate); gnathopod 2 and body of male not setose *Caprella borealis* Mayer, 1903

22a Head with a pair of tubercles 23

22b Head with a pair of rather sharp spines 24

23a Posterior pereonites with large dorsal spines; pereonite 1 of male about the same length as the head *Caprella ferrea* Mayer, 1903

23b Posterior pereonites with small dorsal tubercles; pereonite 1 of male about 1 and one-half times as long as the head *Caprella rudiuscula* Laubitz, 1970

24a Flagellum of antennule longer than the peduncle; pereonites 1 and 2 of male not more than twice as long as wide
Metacaprella anomala (Mayer, 1903)

24b Flagellum of the antennule shorter than the peduncle; pereonites 1 and 2 of male more than twice as long as wide
Metacaprella kennerlyi (Stimpson, 1864)

Dougherty, E. C., & J. Steinberg. 1953. Notes on the skeleton shrimps (Crustacea: Caprellidae) of California. Proc. Biol. Soc. Washington, 66:39-50.

Laubitz, D. 1970. Studies on the Caprellidae (Crustacea, Amphipoda) of the American north Pacific. Publ. Biol. Oceanogr., Mus. Nat. Sci. Canada, vol. vii + 89 pp.

Leung, T. M. 1967. An illustrated key to the species of whale-lice (Amphipoda, Cyamidae), ectoparasites of Cetacea, with a guide to the literature. Crustaceana, 12:279-91.

McCain, J. C. 1968. The Caprellidae (Crustacea: Amphipoda) of the western north Atlantic. Bull. 278, U.S. Nat. Mus. vi + 147 pp.

SUPERORDER EUCARIDA

ORDER EUPHAUSIACEA

These crustaceans are not likely to be collected except by use of large plankton nets towed at depths of 100 m or more (except at night, when certain species migrate closer to the surface). Banner (1950) has provided keys, descriptions, and records of distribution for 13 species found from Alaska to California. The majority of these can be taken off Washington and British Columbia, especially well out from shore, and a few are found regularly in the Strait of Georgia, the region of the San Juan Archipelago, and Puget Sound. The ones most likely to be encountered in the inland waters are *Euphausia pacifica* Hansen, 1911 (without a rostrum) and *Thysanoessa raschii* M. Sars, 1863 (with a slender rostrum).

Banner, A. H. 1950. A taxonomic study of the Mysidacea and Euphausiacea (Crustacea) of the northeastern Pacific. Part III. Euphausiacea. Trans. Royal Canad. Inst., 28(1949):1-63.

ORDER DECAPODA

SUBORDER NATANTIA

1a Third legs chelate; epimera of first abdominal segment overlapping those of the second segment; abdomen without a sharp bend; primarily oceanic, and not likely to be found except in deep water
Section Penaeidea

1b Third legs not chelate; epimera of second abdominal segment overlapping those of the first; abdomen generally with a sharp bend; includes many littoral and shallow-water species
Section Caridea

Section Penaeidea

1a Last 2 pairs of legs well developed
Family Penaeidae

1b Last 2 pairs of legs much reduced, rudimentary, or lacking
Family Sergestidae

Family Penaeidae

Gennadas borealis Rathbun, 1902 is a pelagic species sometimes found in deep plankton tows.

Family Sergestidae

Sergestes atlanticus Bate, 1888 is a pelagic species sometimes found in deep plankton tows.

Section Caridea

1a Legs with exopodites; restricted to deep water 2
1b Legs without exopodites; not necessarily restricted to deep water 3
2a Second maxillipeds lacking exopodites, or exopodites rudimentary; rostrum essentially absent, but there is a sharp tooth that originates slightly posterior to the margin of the carapace Family Pasiphaeidae
2b Second maxillipeds with well-developed exopodites; rostrum prominent, serrated dorsally Family Oplophoridae
3a Both legs of first pair simple Family Pandalidae
3b Both legs of first pair chelate or subchelate 4
4a First pair of legs subchelate; carpus of second pair of legs undivided Family Crangonidae
4b First pair of legs chelate; carpus of second pair of legs subdivided into several articles 5
5a Eyes covered by the carapace; first pair of legs decidedly better developed than the others; rostrum small or absent Family Alpheidae
5b Eyes not covered by the carapace; first pair of legs not decidedly better developed than the others Family Hippolytidae

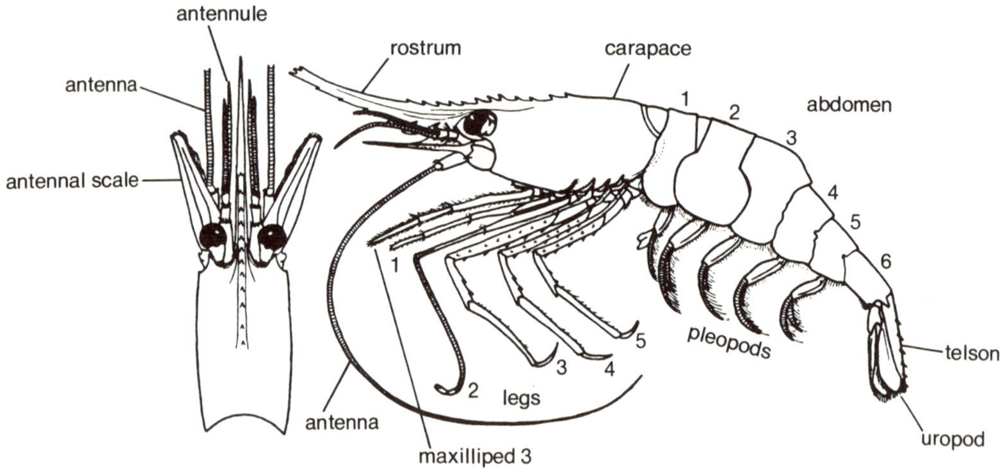

Family Pasiphaeidae

Pasiphaea pacifica Rathbun, 1902 is found in deep water, in inland waters as well as off the coast.

Family Oplophoridae

Hymenodora frontalis Rathbun, 1904 and *Notostomus japonicus* Bate, 1888 are found in deep water off the coast.

Family Pandalidae

1a Antennules about twice as long as the carapace; ischium of first pair of legs with a longitudinal laminate expansion fringed with hairs *Pandalopsis dispar* Rathbun, 1902
(*Pandalopsis ampla* Bate, 1888 is also known to occur in our region, but is restricted to very deep water.)

1b Antennules not longer than the carapace; ischium of first pair of legs without a laminate expansion 2

2a Third segment of abdomen at least partially compressed, with a median dorsal ridge that forms a lobe or spine in front of the posterior margin 3

2b Third segment of abdomen not compressed, without a median dorsal ridge and without a spine in front of the posterior margin 5

3a Third and fourth segments of abdomen with a dorsal median spine on the posterior margin *Pandalus borealis* Krøyer, 1838

3b Third and fourth segments of abdomen without a dorsal median spine on the posterior margin 4

4a Distal half of dorsal edge of rostrum with spines *Pandalus jordani* Rathbun, 1902

4b Distal half of dorsal edge of rostrum without spines *Pandalus goniurus* Stimpson, 1860

5a Dorsal median spines restricted to the anterior half of the carapace (exclusive of the rostrum) 6

5b Dorsal median spines extending posteriorly behind the middle of the carapace (exclusive of the rostrum) 7

6a Sixth abdominal segment about 3 times as long as wide *Pandalus montagui tridens* Rathbun, 1902

6b Sixth abdominal segment less than twice as long as wide *Pandalus platyceros* Brandt, 1851

7a Dorsal spines 17 to 21 *Pandalus hypsinotus* Brandt, 1851

7b Dorsal spines fewer than 15 8

8a Rostrum at least 1 and one half times as long as the rest of the carapace *Pandalus gurneyi* Stimpson, 1871

8b Rostrum less than 1 and one-half times as long as the rest of the carapace 9

9a Telson with 3 to 5 pairs of small lateral spines; distal half of blade of antennal scale narrower than the adjacent thickened axis; outer margin of blade decidedly concave *Pandalus stenolepis* Rathbun, 1902

9b Telson with 6 pairs of small lateral spines; distal half of blade of antennal scale not narrower than the adjacent thickened axis; outer margin of blade nearly straight *Pandalus danae* Stimpson, 1857

Family Crangonidae

1a Rostrum present; eyes not concealed by the carapace as the animal is viewed from above; dactyls of fourth and fifth pairs of legs slender, slightly flattened but not adapted for swimming 5

1b Rostrum absent (although there may be a small rostral spine near the anterior edge of the carapace); eyes nearly concealed by the carapace when the animal is viewed from above; dactyls of fourth and fifth pairs of legs broad and adapted for swimming 2

2a Carapace with 2 median spines in addition to the rostral spine near the anterior margin *Argis dentata* (Rathbun, 1902)

2b Carapace with 3 or 4 median spines in addition to the rostral spine near the anterior margin 3

3a Dorsal ridge on the sixth abdominal segment terminating posteriorly in a small spine *Argis alaskensis* (Kingsley, 1882)

3b Dorsal ridge on the sixth abdominal segment not terminating posteriorly in a small spine 4

4a First to fourth abdominal segments with a dorsal ridge *Argis crassa* (Rathbun, 1899)

4b First to fourth abdominal segments without a dorsal ridge *Argis levior* (Rathbun, 1902)

5a Rostrum about half the length of the rest of the carapace; with only 4 pairs of legs (second pair absent) *Paracrangon echinata* Dana, 1852

5b Rostrum less than half the length of the rest of the carapace; all 5 pairs of legs present 6

6a Tip of rostrum upturned *Sclerocrangon alata* Rathbun, 1902

6b Tip of rostrum not upturned 7

7a Carapace without an appreciable depression occupying the anterior and middle portions of the dorsal surfacw 8

7b Carapace with an appreciable depression occupying the anterior and middle portions of the dorsal surface 15

8a Anterior half of the carapace with 1 or more median spines 9

8b Anterior half of the carapace without any median spines *Crangon stylirostris* Holmes, 1900

9a Anterior half of the carapace with 1 median spine 10

9b Anterior half of the carapace with 2 median spines 14

10a Dorsal surface of sixth abdominal segment with 2 prominent longitudinal ridges *Crangon dalli* Rathbun, 1902

10b Dorsal surface of sixth abdominal segment without longitudinal ridges 11

11a Ventral surface of sixth abdominal segment with a median groove 12

11b Ventral surface of sixth abdominal segment without a median groove *Crangon alba* Holmes, 1900

12a Propodus of first pair of legs at least 4 times as long as wide *Crangon franciscorum* Stimpson, 1856

12b Propodus of first pair of legs less than 3 and one-half times as long as wide 13

13a Anteromesial corner of the blade of the antennal scale a rounded lobe projecting forward about as far as the tip of the anterolateral spine *Crangon nigricauda* Stimpson, 1856

13b Anteromesial corner of antennal scale not projecting forward *Crangon alaskensis* Lockington, 1877

14a Rostrum with a pointed tip and ascending at an angle of about 45°; body surface with fine, short hairs *Crangon communis* Rathbun, 1899

14b Rostrum with a rounded tip and ascending at only a slight angle; body surface without hairs *Crangon resima* Rathbun, 1902

15a Anterior median spine of the carapace projecting beyond the bases of the orbits *Crangon acclivis* Rathbun, 1902

15b Anterior median spine of the carapace barely or not quite reaching the bases of the orbits 16

16a Dorsal surface of the more anterior segments (1 to 5 or 2 to 5) of the abdomen with a median ridge *Crangon variabilis* Rathbun, 1902
16b Dorsal surface of the more anterior segments of the abdomen without a median ridge 17
17a Ridge bordering each side of the dorsal depression of the carapace with a sharp spine *Crangon munitella* Walker, 1898
17b Ridge bordering each side of the dorsal depression of the carapace without a spine *Crangon munita* Dana, 1852

Family Alpheidae

1a Dactyls of legs 3 to 5 slender and not bifid; often found in burrows of the mud shrimp, *Upogebia pugettensis* *Betaeus harrimani* Rathbun, 1904
1b Dactyls of legs 3 to 5 stout and bifid; found under rocks on the open coast *Betaeus setosus* Hart, 1964

Family Hippolytidae

1a Carpus of second pair of legs with 7 articles 2
1b Carpus of second pair of legs with 3 articles *Hippolyte clarki* Chace, 1951
2a Supraorbital spines present on carapace 3
2b Supraorbital spines absent 4
3a Carapace with 2 or more supraorbital spines on each side; third maxilliped with an exopodite 15
3b Carapace with only 1 supraorbital spine on each side; third maxilliped without an exopodite 5
4a Third maxilliped with an exopodite 7
4b Third maxilliped without an exopodite 21
5a Rostrum less than twice as long as the eye, and about half as long as the rest of the carapace; known only from deep water offshore *Lebbeus washingtonianus* (Rathbun, 1902)
5b Rostrum more than twice as long as the eye, and about as long as the rest of the carapace; found in inland waters as well as offshore 6
6a Sides of first to third abdominal segments rounded *Lebbeus grandimanus* (Brashnikov, 1907)
6b Sides of first to third abdominal segments terminating in sharp points *Lebbeus groenlandicus* (Fabricius, 1775)
7a Third, fourth and fifth segments of the abdomen with a dorsal ridge terminating in a sharp spine *Eualus barbatus* (Rathbun, 1899)
7b Third, fourth and fifth segments not all with a dorsal ridge or a sharp spine 8
8a Distal half of the rostrum without dorsal spines *Eualus fabricii* (Krøyer, 1841)
8b At least part of the distal half of the rostrum with dorsal spines 9
9a Distal third of the rostrum without spines ventrally or dorsally; dactyls of legs subchelate *Eualus biunguis* (Rathbun, 1902)
9b Distal third of the rostrum with at least some spines; dactyls of legs not subchelate 10
10a Ventrolateral margin of the fourth abdominal segment with a sharp tooth or spine (this may be small) 11
10b Ventrolateral margin of the fourth abdominal segment without a sharp tooth or spine 14

11a With an epipodite on the first leg, or first and second legs, but not on the third leg *Eualus suckleyi* (Stimpson, 1864)
11b With an epipodite on the first to third legs 12
12a Dorsal margin of the rostrum in the region of the eyes strongly convex *Eualus avinus* (Rathbun, 1899)
12b Dorsal margin of the rostrum in the region of the eyes nearly straight 13
13a Rostrum not reaching forward as far as the second article of the peduncle of the antenna *Eualus pusiolus* (Krøyer, 1841)
13b Rostrum reaching forward as far as the second article of the peduncle of the antenna *E. herdmani* (Walker, 1898)
14a Rostrum about as long or longer than the rest of the carapace; dorsal margin of the rostrum nearly straight or slightly concave *Eualus townsendi* (Rathbun, 1902)
14b Rostrum shorter than the rest of the carapace; dorsal and ventral edges of the rostrum strongly convex *Eualus macrophthalmus* (Rathbun, 1902)
15a Carapace with 3 or 4 supraorbital spines on each side (the anteriormost spine may be much reduced, so that when there are 3 altogether only 2 may be obvious); depth of rostrum more than half its length *Spirontocaris prionota* (Stimpson, 1864)
15b Carapace with 2 supraorbital spines on each side; depth of rostrum not more than half its length 16
16a Posterior margins of first to third abdominal segments forming a rather sharp angle on each side of the body *Spirontocaris lamellicornis* (Dana, 1852)
16b Posterior margins of first to third abdominal segments rounded or forming only a very weak angle on each side of the body 17
17a Epipodites present only on first pair of legs *Spirontocaris sica* Rathbun, 1902
17b Epipodites present on at least first and second pairs of legs 18
18a Epipodites present on first and second pairs of legs 19
18b Epipodites present on first to third pairs of legs 20
19a Rostrum longer than the rest of the carapace, terminating in a slender spine *Spirontocaris holmesi* Holthuis, 1947
19b Rostrum shorter than the rest of the carapace, more or less truncate at the tip *Spirontocaris snyderi* Rathbun, 1902
20a Female: profile of dorsal margin of rostrum and rest of carapace showing a continuous convexity. Male: third abdominal segment, as seen in profile, without a hornlike dorsal process overlapping the fourth segment *Spirontocaris spina* (Sowerby, 1805)
20b Female: profile of dorsal margin of rostrum and rest of carapace not showing a continuous convexity (the dorsal margin of the proximal part of the carapace is almost straight). Male: third abdominal segment, as seen in profile, with a hornlike dorsal process overlapping the fourth segment *Spirontocaris arcuata* Rathbun, 1902
21a Posterior part of third abdominal segment, as seen in profile, with a pronounced dorsal hump 22
21b Posterior part of third abdominal segment, as seen in profile, without a pronounced dorsal hump 24
22a With 3 dorsal teeth on the rostrum and carapace, the base of the anteriormost tooth being at about the same transverse level as the anteriormost surfaces of the eyes *Heptacarpus tridens* (Rathbun, 1902)
22b With 4 or 5 dorsal teeth on the rostrum and carapace, the anteriormost

tooth being decidedly farther forward than the anteriormost surface of the eyes 23
23a Third maxilliped with an epipodite *Heptacarpus flexus* (Rathbun, 1902)
23b Third maxilliped without an epipodite *Heptacarpus tenuissimus* Holmes, 1900
24a Distal half of rostrum without spines *Heptacarpus stylus* (Stimpson, 1864)
24b Distal half of rostrum with spines 25
25a Sixth abdominal segment longer than the telson *Heptacarpus decorus* (Rathbun, 1902)
25b Sixth abdominal segment shorter than the telson 26
26a Maxilliped not reaching beyond the end of the acicle of the antenna *Heptacarpus kincaidi* (Rathbun, 1902)
26b Maxilliped reaching beyond the end of the acicle of the antenna 27
27a Rostrum as long as or slightly longer than the rest of the carapace 28
27b Rostrum shorter than the rest of the carapace 29
28a Dorsal edge of rostrum concave; restricted to deep water *Heptacarpus moseri* (Rathbun, 1902)
28b Dorsal edge of rostrum straight; found in bays, lagoons, and tide pools *Heptacarpus paludicola* Holmes, 1900
29a Rostrum reaching beyond the middle of the antennal scale *Heptacarpus sitchensis* (Brandt, 1851)
29b Rostrum not reaching beyond the middle of the antennal scale 30
30a Rostrum reaching the second article of the antennal peduncle, its dorsal margin only slightly convex in the region of the eyes *Heptacarpus stimpsoni* Holthuis, 1947
30b Rostrum not reaching the second article of the antennal peduncle, its dorsal margin decidedly convex in the region of the eyes 31
31a Tip of telson with 1 or 2 marginal setae *Heptacarpus brevirostris* (Dana, 1852)
31b Tip of telson with 4 or 5 marginal setae *Heptacarpus taylori* (Stimpson, 1857)

SUBORDER REPTANTIA

Section Astacura

Superfamily Thalassinoidea

1a Rostrum conspicuous, about twice as long as wide, with 5 or 6 teeth on each side; found on muddy substrates at depths generally greater than 100 m Family Axiidae
1b Rostrum either a small tooth or a short, broad process that is subdivided into 3 teeth; abundant intertidally in substrates of mud and muddy sand Family Callianassidae

Family Axiidae

Axiopsis spinulicauda (Rathbun, 1902) is occasionally taken by dredging. It occurs on muddy bottoms, usually at depths greater than 100 m.

Family Callianassidae

1a Rostrum distinct, subdivided into 3 teeth, hairy above; first pair of legs almost equal, subchelate (the dactyl bends back to close against the propodus itself, rather than against a clawlike process of the latter) *Upogebia pugettensis* (Dana, 1852)

1b Rostrum reduced to a small tooth, not hairy; first pair of legs decidedly unequal, chelate (the dactyl closes against a clawlike process of the propodus) 2

2a Rostrum bluntly rounded; chela of first pair of legs, when closed, with a conspicuous gap between the dactyl and the clawlike process of the propodus *Callianassa californiensis* Dana, 1854

2b Rostrum sharp; chela of first pair of legs, when closed, with almost no gap between the dactyl and the clawlike process of the propodus *Callianassa gigas* Dana, 1852

Section Anomura

1a Body as a whole almost ovoid; first pair of legs not chelate; living just beneath the surface on sandy beaches exposed to considerable wave action Superfamily Hippoidea: Family Hippidae

1b Body as a whole not ovoid; first pair of legs chelate or subchelate; generally in rocky or stony situations, on muddy bottoms, or living in permanent burrows in mud or muddy sand 2

2a Uropods present, and constituting a well developed tail fan (even in the case of crablike species in which the abdomen is permanently folded against the underside of the thorax) 3

2b Uropods, if present, not constituting a well developed tail fan 4

3a Body somewhat shrimplike, the abdomen bent under itself but not folded against the underside of the thorax Superfamily Galatheoidea: Family Galatheidae

3b Body crablike, the abdomen permanently folded against the underside of the thorax Superfamily Galatheoidea: Family Porcellanidae

4a Uropods present, hooklike, and with a rasplike area on the external surface of each ramus; abdomen soft, unsegmented, not folded against the underside of the thorax, generally twisted (but sometimes straight); living in snail shells, worm tubes, or cavities in sponge masses 5

4b Uropods absent; abdomen either firm or soft, segmented or unsegmented, folded against the underside of the thorax, not twisted; not living in snail shells, worm tubes, or cavities in sponge masses Superfamily Paguroidea: Family Lithodidae

5a Bases of third maxillipeds as close together as those of the first 2 pairs; both sexes with paired pleopods (male with 2 pairs, female with 1 pair); with 13 pairs of gills; chelipeds similar and nearly equal; includes 2 species of hermit crabs found subtidally in our area Superfamily Coenobitoidea: Family Diogenidae

5b Bases of third maxillipeds widely separated; either without paired pleopods, or with paired pleopods only in the male; with 11 pairs of gills; chelipeds generally dissimilar and unequal (includes most hermit crabs found in our area) 6

6a Male without paired pleopods; female with paired gonopores; includes all hermit crabs found intertidally and most subtidal species as well Superfamily Paguroidea: Family Paguridae

6b Male with paired first and second pleopods; female with gonopore on left side only; includes a rare hermit crab of deep water
Superfamily Paguroidea: Family Parapaguridae

Superfamily Hippoidea

Family Hippidae

Emerita analoga (Stimpson, 1857) is occasionally found on sandy beaches of the outer coast of Washington and Vancouver Island. However, the populations are not persistent; they are derived from planktonic larvae originating farther south. *Emerita* is a gregarious animal that moves up and down the beach with the tide.

Superfamily Galatheoidea

Family Galatheidae

1a Eyes with dark pigment, faceted; exopodite of maxilliped terminating in a flagellum; sometimes fairly common at depths greater than about 20 m *Munida quadrispina* Benedict, 1902
1b Eyes without dark pigment, not faceted; exopodite of first maxilliped not terminating in a flagellum; not likely to be taken except in deep water on the continental shelf *Munidopsis quadrata* (Faxon, 1893)

Family Porcellanidae

1a Chelipeds equal or nearly equal, flattened, with a finely granulated surface, not covered with plumose hairs 2
1b Chelipeds decidedly unequal, thick, roughened by tubercles, and covered with plumose hairs 3
2a Carpus of chelipeds about twice as long as wide, with parallel anterior and posterior margins *Petrolisthes eriomerus* Stimpson, 1871
2b Carpus of chelipeds about 1 and one-half times as long as wide, with anterior and posterior margins not parallel
Petrolisthes cinctipes (Randall, 1839)
3a Telson with 7 plates; carpus of chelipeds with 3 or 4 uneven, serrate teeth on the anterior margin *Pachycheles pubescens* Holmes, 1900
3b Telson with 5 plates; carpus of chelipeds with a broad, nearly triangular lobe on the anterior margin *Pachycheles rudis* Stimpson, 1860

Family Lithodidae

1a Abdomen soft, unsegmented, not completely covered by calcareous plates, and not tightly applied to the underside of the thorax 2
1b Abdomen not soft, segmented, covered by calcareous plates, and tightly applied to the underside of the thorax 5
2a Abdomen thin and flat; chelipeds nearly equal (the right cheliped only slightly larger than the left), with the dactyl of each about as long as, or longer than, the propodus
Placetron wosnesenskii Schalfeew, 1894
2b Abdomen thick, chelipeds decidedly unequal, with the dactyl of each shorter than the propodus 3
3a Upper surface of carapace spiny; rostrum tipped by strong spines

Acantholithodes hispidus (Stimpson, 1860)

3b Upper surface of carapace not spiny; rostrum not tipped by sharp spines 4

4a Carapace and legs flattened, very hairy; right cheliped not greatly swollen and not covered with blue granules; anterior margins of second to fourth legs deeply incised (the incisions may be hidden by hairs) *Hapalogaster mertensii* Brandt, 1850

4b Carapace and legs not flattened or very hairy; right cheliped greatly swollen and covered with blue granules; anterior margins of second to fourth legs not deeply incised *Oedignathus inermis* (Stimpson, 1860)

5a Carapace with lateral expansions that completely conceal the legs when the animal is viewed from above 6

5b Carapace without lateral expansions that completely conceal the legs when the animal is viewed from above 7

6a Propodus of chelipeds smooth; rostrum broadest distally; abdominal plates without raised margins *Cryptolithodes sitchensis* Brandt, 1853

6b Propodus of chelipeds with coarse tubercles; rostrum generally broadest proximally; abdominal plates with raised margins *Cryptolithodes typicus* Brandt, 1849

7a Lateral plates of third to fifth abdominal segments with concave central membranous areas; chelipeds and other legs with long, blunt-tipped, flat spines *Phyllolithodes papillosus* Brandt, 1849

7b None of the abdominal plates with concave central membranous areas; chelipeds and other legs without long, blunt-tipped, flat spines (however, tubercles, sharp spines, or hairs may be present) 8

8a Outline of carapace similar to that of an equilateral triangle; upper surface of carapace with a deep semicircular concavity; chelipeds and other legs with spines that are conspicuously hairy *Rhinolithodes wosnessenskii* Brandt, 1849

8b Outline of carapace pentagonal or hexagonal; upper surface of carapace without a deep semicircular concavity; chelipeds and other legs with rounded or spinelike tubercles, but these are not conspicuously hairy 9

9a Tubercles on chelipeds and other legs spinelike; posterior edge of carpus of chelipeds excavated into a deep sinus (when this is in apposition to a shallow sinus on the anterior edge of the carpus of the second leg, a circular hole is formed) *Lopholithodes foraminatus* (Stimpson, 1862)

9b Tubercles on chelipeds and other legs mostly rounded and blunt; posterior edge of carpus of chelipeds not excavated into a sinus *Lopholithodes mandtii* Brandt, 1849

Superfamily Coenobitoidea

Family Diogenidae

1a Eye stalks rather stout, about two-thirds to three-fourths as long as the width of the anterior portion of the carapace *Paguristes turgidus* (Stimpson, 1857)

1b Eye stalks slender, about as long as or slightly longer than the width of the anterior portion of the carapace *Paguristes ulreyi* Schmitt, 1921

Superfamily Paguroidea

Family Paguridae

The key to pagurids is substantially complete for this area. However, the majority of the species are not found intertidally; moreover, of those that are strictly subtidal, some occur only off the open coast.

The following species are regularly found intertidally on the open coast:

Pagurus hirsutiusculus
P. granosimanus
P. beringanus
P. samuelis
P. hemphilli

In inland waters, only the first 3 of the above may be expected.

The species listed below are primarily subtidal, but are occasionally or rarely taken intertidally:

Pagurus caurinus
P. armatus
P. kennerlyi
P. quaylei (open coast)
Discorsopagurus schmitti
Elassochirus gilli

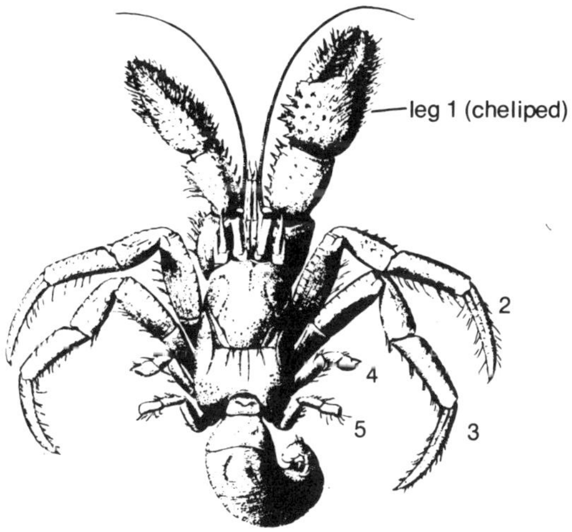

1a Only the shield of the carapace calcified; basal portions of ocular acicles not covered by the carapace, visible from above; fourth legs subchelate; male with pleopods, but these are unpaired 2

1b Nearly all of the carapace calcified; basal portions of ocular acicles covered by the carapace, not visible from above; fourth legs not subchelate; male without pleopods *Labidochirus splendescens* (Owen, 1839)

2a Abdomen not coiled; uropods symmetrical; in shells of scaphopod molluscs or in tubes of sabellariid and serpulid polychaetes (rarely shells of gastropods) 3

2b Abdomen coiled; uropods asymmetrical; in shells of gastropods or in sponge masses 4

3a Palm of right chela wider at the level where the dactyl and fixed claw originate than at its proximal end; right and left halves of telson partially separated by a groove; in shells of scaphopod molluscs; subtidal, off the coast *Orthopagurus minimus* (Holmes, 1900)

3b Palm of right chela little if any wider at the level where the dactyl and fixed claw originate than at its proximal end; right and left halves of telson not separated by a groove; almost always in tubes of sabellariid and serpulid polychaetes (rarely in shells of gastropods); largely subtidal, in inland waters as well as off the coast *Discorsopagurus schmitti* (Stevens, 1925)

4a Left cheliped not appearing to have been twisted to the extent of about 90°; neither the merus nor the carpus of the right cheliped expanded laterally 5

4b Left cheliped appearing to have been twisted to the extent of about 90°; either the merus or carpus of the right cheliped expanded laterally 22

5a Dorsal surface of left chela flattened, and with the dorsolateral and dorsomesial margins of the palm slightly raised; dactyl of second and third legs twisted in relation to the propodus; lateral surface of propodus of second and third legs with 2 prominent reddish brown longitudinal stripes 6

5b Dorsal surface of left chela not flattened, at least much of the palm being convex or even elevated into a plateau, ridge, or crest (but the dorsolateral surface of the palm is concave in certain species); dactyl of second and third legs not twisted in relation to the propodus; propodus of second and third legs without longitudinal stripes 8

6a Dorsal surface of dactyls of second and third legs with a single longitudinal groove; coloration of chelipeds uniform, without dark markings; merus and carpus of chelipeds without an opalescent sheen *Pagurus aleuticus* (Benedict, 1892)

6b Dorsal surface of dactyls of second and third legs with longitudinal rows of small spines or tubercles, these rows separated by 2 shallow grooves; chelipeds with dark markings; merus and carpus of chelipeds with an opalescent sheen 7

7a Mesial surface of palm and dactyl of chelae studded with spines or tubercles; eyes yellow-green; dorsal surface of palm of chelae with a distinct dark band alongside the articulation with the dactyl and continuing along the cutting edge of the fixed claw *Pagurus ochotensis* Brandt, 1851

7b Mesial surface of palm and dactyl of chelae without spines or tubercles except on the upper portion, where an irregular row of spines or tubercles may be present; eyes black; dorsal surface of palm of chelae without a distinct dark band *Pagurus armatus* (Dana, 1851)

8a Dorsal surface of palm of left chela with a prominent ridge or crest near the midline, and with a distinct concavity lateral to this; subtidal; found only in very deep water off the coast 9

8b Dorsal surface of palm or left chela without a prominent ridge or crest near the midline (but there may be a slight longitudinal eminence) and without a concavity; includes intertidal and subtidal species, some of which occur in inland waters 11

9a Right chela about 2 and one-half times as long as wide; the nearly triangular elevation on the dorsal surface of the palm of the right chela continuing onto the fixed claw as a rather sharp, spiny crest *Pagurus tanneri* (Benedict, 1892)

9b Right chela not more than twice as long as broad; the nearly triangular elevation on the dorsal surface of the palm of the right chela continuing onto the fixed claw as a low, spiny or tuberculate ridge 10

10a Apex of the nearly triangular elevation on the dorsal surface of the palm of the right chela forming a prominent spiny or tuberculate projection *Pagurus cornutus* (Benedict, 1892)

10b Apex of the nearly triangular elevation on the dorsal surface of the palm of the right chela not forming a prominent projection *Pagurus confragosus* (Benedict, 1892)

11a Ventral surface of merus of right cheliped with 1 or 2 especially prominent tubercles (these are about twice as large as the other tubercles); includes intertidal and subtidal species 12

11b Ventral surface of merus of right cheliped without any especially prominent tubercles; strictly subtidal species 17

12a Ventral surface of merus of right cheliped with 2 prominent tubercles 13

12b Ventral surface of merus of right cheliped with 1 prominent tubercle 15

13a Ventrolateral surface of dactyl and propodus of left third leg without spines or tubercles, or the propodus with only 1 or 2 spines on its distal portion; legs (including chelipeds) olive green, with very small blue dots *Pagurus granosimanus* (Stimpson, 1858)

13b Ventrolateral surface of dactyl and propodus of left third leg with 1 or more irregular rows of small spines or tubercles; color of legs not as above 14

14a Shield of the carapace about as wide as long; chelae with small spines rather than tubercles or granules; dorsal surface of carpus of left cheliped with a single row of stout spines; rostrum generally rounded; second and third legs with a red band at the articulation of the dactyl and propodus; not restricted to the open coast, but found only in the lowest part of the intertidal region and subtidally *Pagurus beringanus* (Benedict, 1892)

14b Shield of carapace decidedly longer than wide; chelae with tubercles or granules; dorsal surface of carpus of left cheliped with 2 rows of moderately short spines; rostrum acute; dactyl of second and third legs bright blue; restricted to the open coast, where it is found in the mid-intertidal region *Pagurus samuelis* (Stimpson, 1857)

15a Chelae with irregular rows of moderately stout spines, these partially obscured by tufts of long setae; with both solid-colored (reddish brown) antennae and a white band at the articulation of the dactyl and propodus of the second and third legs *Pagurus caurinus* Hart, 1971

15b Chelae with closely-spaced tubercles or granules, these often inconspicuous (scattered tufts of setae may be present or absent); not with both solid-colored antennae and a light band at the articulation of the dactyl and propodus of the second and third legs 16

16a Ventral side of carpus of right cheliped greatly swollen; shield of carapace decidedly longer than wide; antennae orange-red, without white spots; legs on the whole dark red, with minute yellow spots, but without light bands at the articulation of the dactyl and propodus; on the open coast, where it is restricted to the lowest part of the intertidal region and the shallow subtidal region

Pagurus hemphilli (Benedict, 1892)

16b Ventral side of carpus of right cheliped not greatly swollen; shield of carapace about as wide as long, or wider than long; antennae largely green, with white spots; legs greenish brown, the second and third legs with white or bluish white bands at the articulation of the dactyl and propodus; on the open coast and in inland waters, mostly in the mid-intertidal region *Pagurus hirsutiusculus* (Dana, 1851)

17a Dactyl of left chela bowed to the extent that when it is closed against the fixed claw, there is a prominent gap between them; ocular acicles usually terminating in 2 to 5 spines

Pagurus quaylei Hart, 1971

17b Dactyl of left chela not bowed to the extent that when it is closed against the fixed claw, there is a prominent gap between them; ocular acicles terminating in a single spine or merely acute 18

18a Carpus of left cheliped about 3 times as long as wide, and usually appreciably longer than the merus 19

18b Carpus of left cheliped about twice as long as wide, and about the same length as the merus 20

19a Carpus of right cheliped about twice as long as wide; left chela more nearly triangular than elongate-ellipsoidal, and with a double row of divergent spines on the eminence near the midline of the dorsal surface of the palm; rostrum not pronounced, with a blunt tip; distal part of merus of chelipeds without a white band

Pagurus stevensae Hart, 1971

19b Carpus of right cheliped about 1 and one-half times as long as wide; left chela more nearly elongate-ellipsoidal than triangular, and with a single row of large spines on the eminence near the midline of the dorsal surface of the palm; rostrum pronounced (about the shape of an equilateral triangle); distal part of merus of chelipeds with a distinct white band *Pagurus dalli* (Benedict, 1892)

20a Dorsal side of the base of the cornea with a tuft of setae; eyestalks (including the cornea) about 5 times as long as wide; distal portion of merus of cheliped with a white band; antennae with light and dark bands *Pagurus kennerlyi* (Stimpson, 1864)

20b Dorsal side of the base of the cornea without a tuft of setae; eyestalks (including the cornea) generally not more than 4 times as long as wide; distal portion of merus of cheliped without a white band; antennae not banded 21

21a Ventral margin of dactyl of second and third legs with a row of small corneous tubercles; chelae so hairy that the spines or tubercles are nearly obscured; antennae grayish tan; not restricted to the open coast *Pagurus capillatus* (Benedict, 1892)

21b Ventral margin of dactyl of second and third legs with a row of sharp corneous spines; chelae not so hairy that the spines or tubercles are nearly obscured; restricted to the open coast

Pagurus setosus (Benedict, 1892)

22a Carpus of right cheliped not strongly flattened and not expanded into winglike projections; dorsal surface of propodus of right chela roughened by spines and tubercles; lateral and mesial surface of merus of legs (especially chelipeds) bright grayish blue (elsewhere, the legs range from cream to red) *Elassochirus tenuimanus* (Dana, 1851)

22b Carpus of right cheliped strongly flattened and expanded into winglike projections; dorsal surface of propodus of right chela nearly smooth; prevailing coloration of legs orange to red, without grayish blue on the lateral and mesial surface of the merus 23

23a Dorsal surface of carpus of right cheliped with one or more irregular rows of small spines; merus and carpus of chelipeds with lavender-purple areas *Elassochirus cavimanus* (Miers, 1879)

23b Dorsal surface of carpus of right cheliped without small spines; chelipeds rather uniformly orange to red *Elassochirus gilli* (Benedict, 1892)

Family Parapaguridae

Parapagurus pilosimanus benedicti De Saint Laurent, 1972 is taken only in very deep water off the coast.

Section Brachyura

Subsection Brachygnatha

1a Carapace narrowed anteriorly into a rostrum consisting of 2 pointed "horns" (these may be joined medially); eyes usually not situated in orbits Infrasubsection Oxyrhyncha

1b Carapace not narrowed anteriorly into a distinct rostrum; eyes situated in orbits Infrasubsection Brachyrhyncha

Infrasubsection Oxyrhyncha

1a Rostrum about half the length of the rest of the carapace, consisting largely of 2 slender, spinelike processes 2

1b Rostrum not more than one-third or two-fifths the length of the rest of the carapace, consisting of 2 broad, flattened processes 3

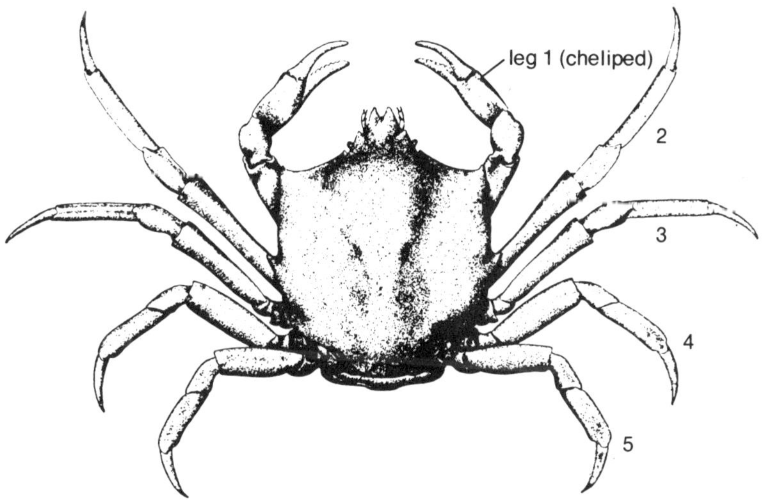

2a Spinelike processes of rostrum separate for nearly the entire length of the latter and nearly parallel; dorsal surface of carapace with a large, sharp tooth behind each eye, but otherwise rather smooth *Oregonia gracilis* Dana, 1851

2b Spinelike processes of rostrum usually about two-thirds the length of the latter as a whole, and usually diverging from one another; dorsal surface of carapace without a large, sharp tooth behind the eye, but otherwise rather spiny *Chorilia longipes* Dana, 1851

3a Carapace (including the rostrum) about as wide or wider than long 4

3b Carapace (including the rostrum) decidedly longer than wide 6

4a Carapace, viewed from above, nearly hexagonal; lateral margins of carapace extended into thin winglike sheets, each with 2 sharp-tipped lobes; dorsal surface of carapace nearly smooth; intertidal as well as subtidal *Mimulus foliatus* Stimpson, 1860

4b Carapace, viewed from above, broadly oval or nearly circular; lateral margins of carapace not extended into winglike sheets; dorsal surface of carapace with tubercles or spines, or both; strictly subtidal 5

5a Rostrum not sharply inclined upward; dorsal surface of carapace roughened by tubercles, but not especially spiny; at depths as shallow as 20 m *Chionoecetes bairdi* Rathbun, 1893

5b Rostrum inclined sharply upward; dorsal surface of carapace spiny; generally at depths greater than 500 m *Chionoecetes angulatus* Rathbun, 1893

6a Carapace with a prominent outward-directed projection on each side near or behind the middle of the body 7

6b Carapace without a prominent outward-directed projection on each side near or behind the middle of the body (however, there may be an anteriorly directed projection near the middle of the body) 9

7a Dorsal surface of carapace very smooth; distance between the eyes less than about one-third the greatest width of the carapace *Pugettia producta* Randall, 1839

7b Dorsal surface of carapace with tubercles; distance between the eyes equal to about one-half the greatest width of the carapace 8

8a Posteriormost projections of the lateral margins of the carapace extending outward farther than those close to the middle, thus marking the widest part of the carapace; tubercles on the dorsal surface all about the same size *Pugettia richii* Dana, 1851

8b Posteriormost projections of the lateral margins of the carapace not extending outward farther than those close to the middle; tubercles on the dorsal surface of different sizes *Pugettia gracilis* Dana, 1851

9a Outline of body, as viewed from above, lyre-shaped, owing to a hard, toothed expansion of the carapace on both sides of the anterior half; rostral horns widest at the base *Hyas lyratus* Dana, 1851

9b Outline of body, as viewed from above, more nearly triangular than lyre-shaped, becoming narrowed gradually toward the eyes; rostral horns widest near the middle *Scyra acutifrons* Dana, 1851

Infrasubsection Brachyrhyncha

1a Outline of carapace nearly square, with the lateral margins almost parallel, though converging slightly posteriorly 2

1b Outline of carapace circular, oblong, pentagonal, hexagonal, or some other shape, but not nearly square and not with almost parallel lateral margins 3

2a Legs fringed with abundant hairs; prevailing coloration generally grayish green; chelipeds without conspicuous purple spots *Hemigrapsus oregonensis* (Dana, 1851)

2b Legs not hairy; prevailing coloration generally reddish; chelipeds with conspicuous purple spots *Hemigrapsus nudus* (Dana, 1851)

A crab somewhat similar to both of the above is *Pachygrapsus crassipes* Randall, 1839, found on the coast of Oregon and southward. It is typically blackish green, with red or purple transverse lines on the dorsal surface, and lives at rather high tide levels in rocky intertidal areas and salt marshes. Two oceanic species of the genus *Planes* (*P. marinus, P. cyaneus*), closely related to *Pachygrapsus*, are occasionally washed in with clumps of pelagic goose barnacles or seaweeds. (See Chace, 1951.)

3a Margin of the carapace between the eyes with distinct teeth 4

3b Margin of the carapace between the eyes without teeth 9

4a Margin of the carapace between the eyes with 6 conspicuous teeth, the 4 teeth between the antennae smaller than the 2 outer ones; teeth of lateral margins of carapace spiny; carapace and legs covered with stiff hairs; prevailing coloration generally yellowish or greenish yellow *Telmessus cheiragonus* (Tilesius, 1815)

4b Margin of carapace between the eyes with 5 teeth; teeth of lateral margins of carapace not spiny; carapace and legs either not hairy or with soft hairs; prevailing coloration almost never yellowish or greenish 5

5a Tips of dactyl and fixed claw of chela dark, nearly black 6

5b Tips of dactyl and fixed claw of chela not dark 8

6a Chelipeds and dorsal surface of carapace distinctly hairy *Cancer branneri* Rathbun, 1898

6b Chelipeds and dorsal surface of carapace not hairy, though either or both may be roughened by tubercles 7

7a Carapace not more than 1 and one-half times as wide as long; width of carapace not often exceeding 4 cm; generally nestling in holes in rock or wood, or in empty shells of large barnacles, etc. *Cancer oregonensis* (Dana, 1852)

7b Carapace more than 1 and one-half times as wide as long; width of carapace frequently attaining 15 cm; in sandy and muddy bays, especially where there is eelgrass, as well as on rocky shores *Cancer productus* Randall, 1839

8a Carapace widest at the level of the tenth in the series of teeth lateral to the eyes on either side of the body; prevailing coloration of dorsal surface of the carapace light reddish brown or grayish brown; width of carapace frequently exceeding 15 cm *Cancer magister* Dana, 1852

8b Carapace widest at the level of the ninth in the series of teeth lateral to the eyes on either side of the body; prevailing coloration of dorsal surface of the carapace generally olive, with small reddish brown spots; width of carapace up to about 9 cm *Cancer gracilis* Dana, 1852

9a General coloration ranging from gray to nearly red; claws of chela very dark; carapace hard; anterior margin of carapace, between its widest points, with several prominent teeth; free-living *Lophopanopeus bellus* (Stimpson, 1860)

9b General color usually ivory, brownish ivory, or grayish ivory, definitely never nearly red; claws of chela not dark; carapace of certain species, or at least 1 sex of certain species, soft; anterior margin of carapace, between its widest points, without any teeth; generally commensal with other animals (in tubes, burrows, mantle cavities, etc.) 10

10a Carapace more than 1 and one-half times as wide as long 11

10b Carapace not more than 1 and one-half times as wide as long 16
11a Tips of dactyls of the fourth (next to last) pair of legs strongly curved 12
11b Tips of dactyls of the fourth pair of legs not curved 13
12a Dactyl and fixed claw of the chela curved to such an extent that there is a gap between them when they are touching (in bivalve molluscs) *Pinnixa littoralis* Holmes, 1895
12b Dactyl and fixed claw of the chela not curved, closing tightly against one another (in bivalve molluscs) *Pinnixa faba* (Dana, 1851)
13a Dactyls of fourth and fifth pairs of legs shorter than those of the second and third pairs (in tubes of terebellid polychaetes) *Pinnixa tubicola* Holmes, 1895
13b Dactyls of fourth and fifth pairs of legs about equal in length to those of the second and third pairs 14
14a Dactyls of fourth and fifth pairs of legs cylindrical *Pinnixa occidentalis* Rathbun, 1893
14b Dactyls of fourth and fifth pairs of legs distinctly flattened 15
15a Inner margin of dactyl of chela with a low tooth; dactyls of the third pair of legs not as long as those of the fourth pair (in burrows of *Upogebia*) *Pinnixa schmitti* Rathbun, 1904
15b Inner margin of dactyl of chela without a tooth; dactyls of third pair of legs as long or longer than those of the fourth pair (in burrows of *Abarenicola*) *Pinnixa eburna* Wells, 1928
16a Carapace nearly 1 and one-half times as wide as long (in burrows of *Upogebia* and *Callianassa*) *Scleroplax granulata* Rathbun, 1893
16b Carapace about as long as wide 17
17a Carapace with a groove extending backward on the dorsal surface from the orbit of each eye; third pair of legs decidedly the longest (in bivalve molluscs, ascidians) *Fabia subquadrata* Dana, 1851
17b Carapace without a groove extending backward on the dorsal surface from the orbit of each eye 18
18a Dactyl of fifth leg decidedly longer than dactyls of second to fourth legs; dorsal surface of carapace with a pair of tubercles just behind the middle (in ascidians) *Pinnotheres taylori* Rathbun, 1918
18b Dactyl of fifth leg about the same length as dactyls of second to fourth legs; dorsal surface of carapace without a pair of tubercles just behind the middle (in ascidians) *Pinnotheres pugettensis* Holmes, 1900

Chace, F. A., Jr. 1951. The oceanic crabs of the genera *Planes* and *Pachygrapsus*. Proc. U. S. Nat. Mus., 101:65-103.

------. 1951. The grass shrimps of the genus *Hippolyte* from the west coast of North America. J. Wash. Acad. Sci., 41:35-9.

Garth, J. S. 1958. Brachyura of the Pacific coast of America. Oxyrhyncha. Allan Hancock Found. Pacific Exped., 21:1-854.

Hart, J. F. L. 1937. Larval and adult stages of British Columbia Anomura. Canad. J. Res., D, 15:179-220.

------. 1964. Shrimps of the genus *Betaeus* on the Pacific coast of North America with description of three new species. Proc. U. S. Nat. Mus., 115:433-66.

------. 1971. New distribution records of reptant decapod Crustacea, including descriptions of three new species of *Pagurus*, from the waters adjacent to British Columbia. J. Fish. Res. Bd. Canada, 28:1527-44.

Holthuis, L. B. 1947. The Decapoda of the Siboga Expedition. Part IX. The

Hippolytidae and Rhynchocinetidae collected by the Siboga and Snellius Expeditions with remarks on other species. Siboga-Expeditie, 39a[8] (livr. 140). 100 pp.

------. 1955. The recent genera of the caridean and stenopodidean shrimps (Class Crustacea, Order Decapoda, Supersection Natantia) with keys for their determination. Zool. Verhandel. Leiden, no. 26. 157 pp.

McLaughlin, P. A. 1974. The hermit crabs (Crustacea Decapoda, Paguridea) of northwestern North America. Zool. Verhandel. Leiden, no. 130. 396 pp.

Makarov, V. V. 1938. [Crustacea. Anomura]. *In* Fauna SSSR, n.s., no. 16, 10. x + 324 pp. Moskva and Leningrad: Akademii͡a Nauk. (in Russian) (English translation, 1962. Jerusalem: Israel Program for Scientific Translations.)

Man, J. G. de. 1920. The Decapoda of the Siboga Expedition. Part IV. Families Pasiphaeidae, Stylodactylidae, Hoplophoridae, Nematocarcinidae, Thalassocaridae, Pandalidae, Psalidopodidae, Gnathophyllidae, Processidae, Glyphocrangonidae and Crangonidae. Siboga-Expeditie, 39a[3] (livr. 87). 318 pp.

Rathbun, M. J. 1904. Decapod crustaceans of the northwest coast of North America. Harriman Alaska Exped., 10:1-210.

------. 1918. The grapsoid crabs of America. Bull. 97, U. S. Nat. Mus. xxii + 461 pp.

------. 1925. The spider crabs of America. Bull. 129, U. S. Nat. Mus. xx + 613 pp.

------. 1930. The cancroid crabs of America of the families Euryalidae, Portunidae, Atelecyclidae, Cancridae, and Xanthidae. Bull. 152, U. S. Nat. Mus. xvi + 609 pp.

------. 1937. The oxystomatous and allied crabs of America. Bull. 166, U. S. Nat. Mus. vi + 278 pp.

Schmitt, W. L. 1921. The marine decapod Crustacea of California. Univ. Calif. Publ. Zool., 23:1-470.

Stevens, B. A. 1925. Hermit crabs of Friday Harbor, Washington. Publ. Puget Sound Biol. Station, 3:273-309.

------. 1928. Callianassidae from the west coast of North America. Publ. Puget Sound Biol. Station, 6:315-69.

Stevens, B. A., & F. A. Chace, Jr. 1965. The mesopelagic caridean shrimp *Notostomus japonicus* Bate in the northeastern Pacific. Crustaceana, 8:277-84.

Wells, W. W. 1928. Pinnotheridae of Puget Sound. Publ. Puget Sound Biol. Station, 6:283-314.

------. 1940. Ecological studies on the pinnotherid crabs of Puget Sound. Univ. Wash. Publ. Oceanogr., 2:19-50.

CLASS INSECTA

Relatively few insects are found in habitats that are truly marine. However, representatives of several families of Diptera, Hemiptera, and Coleoptera do occur in our region, either in rocky intertidal areas or in pools of high salinity. Where the water is brackish, additional types, especially dipterans, may be abundant. At the fringes of the sea--as under driftwood and decaying seaweed high on sandy beaches or in salt marshes--there is an interesting variety of insects. *Light's Manual* has a good account of the insect fauna found in marine and maritime situations in central California, and an extensive bibliography for the Pacific coast. The few references cited below should also be helpful with respect to our region.

GENERAL

Essig, E. O. 1958. Insects and Mites of Western North America. 2nd ed. New York: Macmillan Co. 1050 pp.

Saunders, L. G. 1928. Some marine insects of the Pacific coast of Canada. Ann. Entom. Soc. Amer., 21:521-45.

Usinger, R. L. (ed.). 1956. Aquatic Insects of California. Berkeley and Los Angeles: University of California Press. ix + 508 pp.

ORDER COLEOPTERA

Blackwelder, R. E. 1931. The genus *Endeodes* Le Conte (Coleoptera, Melampyridae). Pan-Pacific Entomol., 8:128-36.

Hatch, M. H. 1957. The Beetles of the Pacific Northwest. Part II. Staphyliniformia. Seattle: University of Washington Press. ix + 384 pp.

------. 1965. The Beetles of the Pacific Northwest. Part IV. Macrodactyles, Palpicornes, and Heteromera. Seattle: University of Washington Press. viii + 268 pp.

Moore, I. 1954. Notes on *Endeodes* Le Conte with a description of a new species from Baja California. Pan-Pacific Entomol., 30:195-8.

------. 1956. A revision of the Pacific coast Phytosi with a review of the foreign genera (Coleoptera: Staphylinidae). Trans. San Diego Soc. Nat. Hist., 12:103-52.

Spilman, T. J. 1967. The heteromerous intertidal beetles. Pacific Insects, 9:1-21.

ORDER COLLEMBOLA

Dexter, R. W. 1943. *Anurida maritima:* an important sea-shore scavenger. J. Econ. Entomol., 36:797.

ORDER DIPTERA

Cole, F. R. 1969. The Flies of Western North America (with collaboration of E. I. Schlinger). Berkeley and Los Angeles: University of California Press. 693 pp.

Morley, R. L., & R. A. Ring. 1972a. The intertidal Chironomidae (Diptera) of British Columbia I. Keys to their life stages. Canad. Entomol., 104:1093-98.

------. 1972b. The intertidal Chironomidae of British Columbia II. Life history and population dynamics. Canad. Entomol., 104:1099-1121.

Saunders, L. G. 1928. Some marine insects of the Pacific coast of Canada. Ann. Entomol. Soc. Amer., 21:521-45.

ORDER HEMIPTERA

Herring, J. L. 1961. The genus *Halobates* (Hemiptera: Gerridae). Pacific Insects, 3:223-305.

Hutchinson, G. E. 1931. On the occurrence of *Trichocorixa* Kirkaldy (Corixidae, Hemiptera-Heteroptera) in salt marshes and its zoogeographical significance. Amer. Nat., 65:573-4.

ORDER THYSANURA

Benedetti, R. 1973. Notes on the biology of *Nemachilis halophila* in a California sandy beach. Pan Pacific Entomol., 49:246-9.

PHYLUM PHORONIDA

1a With a distinct groove and collar at the base of the lophophore; length regularly exceeding 6 cm; tubes typically buried in sand or in a mixture of sand and gravel *Phoronopsis harmeri* Pixell, 1912

1b Without a distinct groove and collar at the base of the lophophore; length not exceeding 5 cm; either burrowing in calcareous substrates (including shells) or attached to rock or wood 2

2a Length only about 5 mm; burrowing in shells of barnacles and molluscs (especially *Balanus nubilus, Pododesmus cepio,* and *Hinnites giganteus)* and occasionally in limestone; mostly subtidal *Phoronis ovalis* T. S. Wright, 1856

2b Length usually between 2 cm and 5 cm; typically in clumps of several to many specimens attached to rock (especially sandstone and limestone) or wood, the tubes partially covered with silt; found intertidally and on floats *Phoronis ijimai* Oka, 1897

Emig, C. C. 1971. Remarques sur la systématique des Phoronidea X. Notes sur l'écologie, la morphologie et la taxonomie de *Phoronis ijimai* et de *P. vancouverensis*. Marine Biology, 8:154-9.

Marsden, J. R. 1959. Phoronidea from the Pacific coast of North America. Canad. J. Zool., 37:87-111.

Pixell, H. L. M. 1912. Two new species of the Phoronidea from Vancouver Island. Q. J. Micr. Sci., 58:257-84.

Zimmer, R. L. 1967. The morphology and function of accessory reproductive glands in the lophophores of *Phoronis vancouverensis* and *Phoronopsis harmeri*. J. Morph., 121:159-78.

PHYLUM BRYOZOA (ECTOPROCTA)

The bryozoans are among the more difficult groups for a non-specialist to deal with. A very elaborate terminology has been developed by experts to describe species, genera, and higher taxa. Superficial appearances are of limited usefulness in schemes of classification that are based, insofar as possible, on natural affinities. The principal monographs dealing with the rich bryozoan fauna of our area generally either do not provide practical keys to the higher categories, or the keys are not sufficiently unequivocal to be of real help. There are so many exceptions that it is impossible to characterize the higher catagories in only a few words. Moreover, some of the groups have not yet been stabilized to the satisfaction of experts. In general, it is probably fair to say that most of the keys that have been developed to cover extensive bryozoan faunas are not likely to be used by the experts who wrote them, and they do not function acceptably in the hands of non-specialists.

The most useful single work for identification of Bryozoa of our area is that of Osburn (1950, 1952, 1953). It is well illustrated, especially as far as details of zooecial structure are concerned. Unfortunately, it does not have enough illustrations of colony forms. The work of Kli͡uge (1962) on bryozoans of northern seas of the U.S.S.R. is rather good for illustrations of colonies, and because many of the genera and even species in the region covered by Kli͡uge are represented in our area, his monograph is helpful here. Unfortunately, the names for species, genera, and higher categories as used by Kli͡uge are not consistent with those employed by Osburn, so one must do some scratching around to reconcile taxa that are identical or nearly so. The older works of Robertson (1905, 1908, 1910), O'Donoghue (1925) and O'Donoghue & O'Donoghue (1923, 1925, 1926) may be helpful with certain groups, at least, but it must be pointed out that much has happened in systematics of Bryozoa since they were published. A recent paper of Ross (1970) characterizes most of the taxa of cyclostomatous bryozoans likely to be found in our region, and has numerous illustrations in proportion to the relatively few genera and species involved.

In preparing the keys presented here, an effort was made to use superficial characteristics of the colony as much as possible, and to simplify the terminology. The names of genera and species conform to those employed by Osburn. As a rule, when a genus is represented in our area by just 1 species, or by 2 very different species, these are indicated. If, however, there are 2 or more rather similar species, only the genus name is given. One must then turn to the appropriate section of Osburn's monograph. The inconvenience of having to consult another work is offset by an obvious advantage: it will enable the person using this key to confirm his identification of the genus. In fact, it is advisable to check in this way almost every tentative identification made with the help of the simplified key.

There are about 75 genera of Bryozoa in our area, the majority of which fall into the order Cheilostomata. The key covers most of them; only those not likely to be encountered in the course of shore collecting and in dredging at depths of less than about 200 m are omitted. It is to be hoped that if any of these are found, they will not go through the keys so readily as to be misidentified.

CLASS GYMNOLAEMATA

1a Zooecia chitinous, not calcified, often united into a fleshy mass; aperture bordered by a closing apparatus consisting of a collarlike membrane with several to many longitudinal folds — Order Ctenostomata

1b Zooecia usually calcified to at least some extent, never constituting a fleshy mass; aperture usually either provided with a lidlike operculum or closed simply by the membrane that extends from the wall of the zooecium to the portion of the zooid that withdraws (in any case, the aperture is not bordered by a collarlike membrane with longitudinal folds) — 2

2a Zooecia usually tubelike, the aperture closed simply by the membrane that extends from the wall of the zooecium to the portion of the zooid that withdraws; aperture circular; spines rarely present near the aperture or elsewhere; avicularia and vibracula absent — Order Cyclostomata

2b Zooecia rarely tubelike, usually more nearly the shape of an elongated box (rectangular, hexagonal, oval, or some other shape), with the aperture on the frontal surface; aperture of varying shape (circular, semicircular, oval, etc.), often with 1 or more notches, a shelflike

projection, or some other elaboration, and generally closed by a lid-like operculum; spines frequently present near the aperture or elsewhere; avicularia and/or vibracula frequently present
Order Cheilostomata

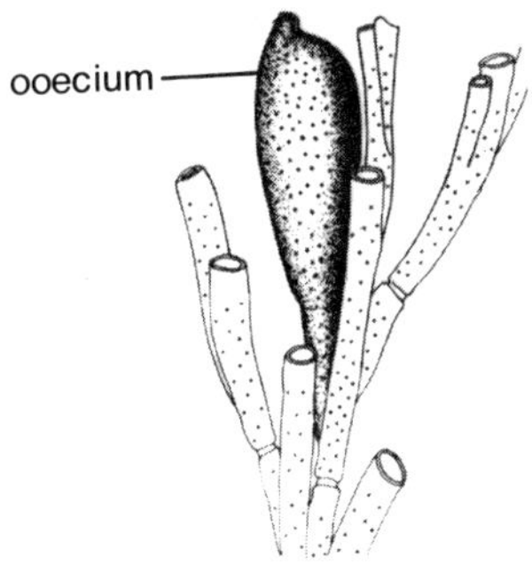

Filicrisia
ORDER CYCLOSTOMATA

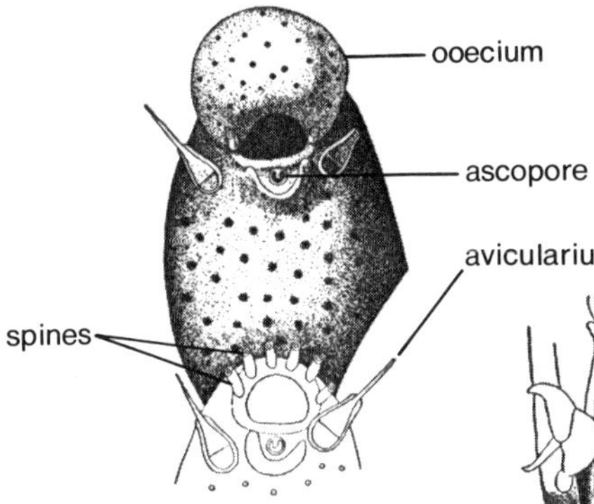

Microporella

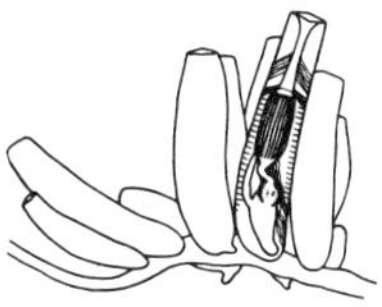

Bowerbankia
ORDER CTENOSTOMATA

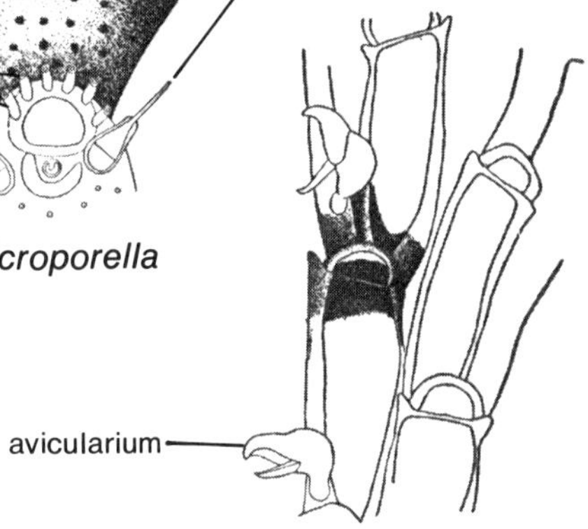

Bugula
ORDER CHEILOSTOMATA

ORDER CTENOSTOMATA

1a Zooids united into a compact, fleshy or leathery mass, sometimes encrusting, sometimes erect and lobed 2

1b Zooids not united into a compact, fleshy or leathery mass, but arising simply from creeping stolons 4

2a Colony club-shaped, consisting of a ringed, muscular stalk composed of zooecia of specialized zooids (most of which are no longer alive) and a capitulum of feeding zooids (the stalk is 2 to 5 times the length of the capitulum, and the colony as a whole may be more than 5 cm high)
Clavopora occidentalis (Fewkes, 1889)

2b Colony encrusting or erect, but of similar consistency throughout, not differentiated into a ringed, muscular stalk and a capitulum (however, the colony may be several cm tall and a short basal portion of it may be stalklike) 3

3a Colony covered by branched or unbranched chitinous spines; zooecial aperture more nearly 2-lipped than circular, color usually yellowish brown *Flustrella*

3b Colony without spines; zooecial aperture circular, color usually whitish or yellowish gray *Alcyonidium*

4a Zooids arising from a basal stolon, this characterized by conspicuous internal septa 5

4b Zooids arising from 1 or more stolonlike outgrowths (these sometimes with delicate septa) of the basal portions of other zooids
Nolella stipata Gosse, 1855

5a Zooids on slender stalks *Triticella*
5b Zooids sessile on the stolon 6
6a Zooids standing erect with respect to the stolon *Bowerbankia*
6b One side of each zooid applied for much of its length to the stolon *Buskia nitens* Alder, 1857

ORDER CYCLOSTOMATA

1a Colony divided, by constrictions at more or less regular intervals, into internodes ("segments") 2
1b Colony not divided into internodes 5
2a Sterile internodes (those without ooecia) consisting of several to many zooecia arranged in a double series *Crisia*
2b Sterile internodes consisting of 1 to several zooecia arranged in a single series 3
3a At least many of the zooecia provided with long, filiform spines 4
3b None of the zooecia provided with filiform spines *Filicrisia*
4a Sterile internodes consisting of a single zooecium; aperture of ooecium terminal *Crisidia cornuta* (Linnaeus, 1758)
4b Sterile internodes generally consisting of 2 zooecia, sometimes of only 1 zooecium; aperture of ooecium at 1 side of the latter, not terminal *Bicrisia edwardsiana* (d'Orbigny, 1839)
5a Colony a short-stalked cup or disk; zooecia arranged in a number of bundles radiating outward from near the edge of the cup or disk (about 12 to 20 zooecia in each bundle, fused together for their entire length) *Discocytis canadensis* O'Donoghue & O'Donoghue, 1926
5b Colony not a stalked cup or disk; zooecia not arranged in bundles radiating outward from near the edge of the disk 6
6a Colony resembling a coral, consisting of heavily-calcified cylindrical branches up to about 5 mm in diameter; zooecia of 2 types (larger ones containing functional zooids, and smaller ones without zooids), these about equally numerous and almost evenly distributed; ooecia not visible from the surface, though their position is sometimes marked by a slight swelling 7
6b Colony encrusting or erect, but if erect, not forming cylindrical branches; zooecia of 2 types, but these not equally numerous and not evenly distributed; ooecia visible 8
7a Branches frequently anastomosing; larger zooecia generally protruding well above the surface of the colony; apertures of most of the smaller zooecia closed *Heteropora pacifica* Borg, 1933
7b Branches of colony anastomosing rarely or not at all; larger zooecia protruding very slightly above the surface of the colony; apertures of most of the smaller zooecia open, except in older portions of the colony *Heteropora magna* O'Donoghue & O'Donoghue, 1923
H. magna and *H. pacifica* may be identical, and both may be synonymous with *H. neozelanica* Busk, 1879.
8a Colony encrusting, usually either discoid, wartlike, or hemispherical, but sometimes forming a nearly nodular growth by encircling a slender object; zooecia of functional zooids elevated, and often in distinct single, double, or multiple rows radiating from near the center of the colony; zooecia of non-functional zooids occupying most of the surface between those of the functional zooids 9
8b Colony encrusting and irregularly lobed, or erect and branching, the branches at least somewhat flattened; zooecia of functional zooids el-

evated, and sometimes in rows having various orientations, but not clearly radiating from near a central point (when the colony is branched, the rows of zooecia are often arranged nearly transversely with respect to the long axis of the branch; their elevated portions are generally restricted to 1 side of the branch); non-functional zooids present, but without apertures, so not readily apparent 10

9a At least many of the pores (cancelli) between functional zooids thin-walled, covered over by a perforated, calcified membrane *Lichenipora*

9b Pores between functional zooids thick-walled, not covered over by a perforated membrane (but some of them may be partially closed by thickening of their rims) *Disporella*

10a Colony generally encrusting, or with peripheral portions becoming erect, but sometimes almost entirely erect; zooecial apertures only on the upper side (or on the side of an erect portion that is continuous with the upper surface of the encrusting portion) 11

10b Colony erect and branching; zooecial apertures in transverse rows on opposite sides of the branches (the ooecia are located in the space between the rows of 1 side and those of the other) *Idmonea*

11a Colony encrusting, forming ribbonlike or fanlike configurations; ooecia compact 12

11b Colony generally encrusting and usually lobed or fanlike, but sometimes with erect, rather narrow branches; ooecia spreading out between many zooecial tubes 15

12a Zooecia, on the whole, separated, not forming distinct bundles or rows in which the zooecia are joined nearly or quite to their apertures (but many of the zooecia may be in groups, and a few may be contiguous) 13

12b Zooecia, on the whole, in bundles or tight groups of 2 to 8, joined nearly or quite to their apertures (a few of the zooecia may be well-separated, however) *Filifascigera fasciculata* (Hincks, 1880)

13a Branches of colony mostly consisting of single series of zooecia (except where an ooecium is present, in which case the zooecia are on both sides of it) *Stomatopora granulata* (Milne-Edwards, 1838)

13b Branches of colony not consisting of single series of zooecia (except perhaps in a restricted portion of the colony) 14

14a Colony an irregular disk or divided into irregular, fan-shaped lobes *Oncousoecia diastoporoides* (Norman, 1868)

14b Colony consisting of strap-shaped or rather narrow branches *Proboscina*

15a Colony almost always encrusting; zooecia not in distinct bundles, clusters, or series 16

15b Colony usually encrusting and with fan-shaped lobes, but may be erect and branched; zooecia usually in bundles, clusters, or series *Tubulipora*

16a Zooecia of 2 sizes, the smaller ones interspersed among the larger ones *Diplosolen obelium* (Johnston, 1838)

16b Zooecia all of about the same size 17

17a Aperture of ooecium situated on its distal margin (the margin that is nearest the growing part of the branch or lobe) *Plagioecia*

17b Aperture of ooecium generally situated some distance from the distal margin, usually near the middle (this is because the ooecium continues to expand distally after its aperture has been differentiated) *Diaperoecia*

ORDER CHEILOSTOMATA

1a Colonies creeping, the branches composed of separate single rows of zooids (in some species of *Beania,* the colony may be organized as a network) 2

1b Colonies forming either thin crusts that adhere tightly to the substrate, sheets that are largely or partially free of the substrate, flexible, erect, branching growths (the few genera that include species composed of single rows of zooids are always flexible, erect growths), or hard erect growths (usually branching) 3

2a Each zooid partly adherent, partly erect; the erect portion (which may be finely ringed, and which may bend near its middle) of nearly equal thickness throughout, with a membranous area on 1 side, just proximal to the terminal aperture and operculum; without spines *Aetea*

2b Zooids without a distinctly erect portion; zooids divided into 2 decidedly unequal parts, the proximal portion slender and tubular, the distal portion inflated and somewhat podlike, with the frontal membrane occupying most of the frontal surface; with small spines on both sides of the frontal membrane, and also with a few spines on the back of the zooecia (one of these may become modified into a holdfast) *Beania*

3a Frontal wall of zooecia largely calcified, though it may be porous or latticelike 4

3b Frontal wall of zooecia membranous, not calcified except perhaps at the edges 31

4a Frontal wall with distinct pores (these may be restricted to marginal areas) 5

4b Frontal wall not porous (however, it may have riblike thickenings with grooves between them, or it may simply be marked with shallow grooves) 30

5a Pores of frontal walls in rather distinct rows, these having an approximately transverse or almost radial orientation (the pores are remnants of uncalcified areas between spines that grow inward and that otherwise become almost completely fused) 6

5b Pores of frontal wall not in distinct rows having an approximately transverse or nearly radial orientation 10

6a Avicularia large, spatulate; ooecia absent *Lyrula hippocrepis* (Hincks, 1882)

6b Avicularia, if present, small and not spatulate; ooecia present 7

7a Avicularia present 8

7b Avicularia absent 9

8a With a single bristlelike avicularium on both sides of the aperture, at about the level of its proximal margin; riblike thickenings and pores absent from the area near the proximal margin of the aperture *Puellina setosa* (Waters, 1899)

8b Avicularia not bristlelike (the mandible is of moderate size), scattered between the zooecia; riblike thickenings and pores present in the area near the proximal margin of the aperture *Colletosia radiata* (Moll, 1803)

9a Ooecia large, prominent, almost or quite half the size of the zooecia *Reginella*

9b Ooecia small, inconspicuous, much less than half the size of the zooecia *Cribrilina*

10a Nearly all (and usually all) of the pores arranged in a single series near the edges of the frontal wall 11

10b Pores scattered over a considerable area of the frontal wall, and frequently rather evenly distributed over most of its surface 16

11a Without any avicularia; with a shelflike elaboration (lyrula) of the proximal margin of the aperture; with a conspicuous elevation (mucro) near the frontal margin of the aperture; a few long spines present next to the distal and lateral margins of the aperture *Mucronella*

11b With avicularia on the frontal wall (near to or remote from the aperture) and sometimes also between the zooecia; a shelflike elaboration (lyrula) of the proximal margin of the aperture may be present or absent; if an elevation is present near the proximal margin of the aperture, this bears an avicularium at its summit or on 1 slope; long spines may be present next to the aperture, or absent 12

12a With an avicularium lateral to the aperture on 1 or both sides, or very close to the proximal margin of the aperture (there may be another avicularium remote from the aperture) 13

12b Without avicularia lateral to the aperture or very close to the frontal margin of the aperture (the avicularium is decidedly away from the aperture) *Parasmittina*

13a With an avicularium lateral to the aperture on 1 or both sides 14

13b With an avicularium very close to the proximal margin of the aperture, often on an eminence, either on the midline or to 1 side of the midline 15

14a Aperture generally almost circular, its diameter nearly or fully half the length of the zooecium; shape of zooecia variable, often irregular, frequently as wide as long; ooecia absent *Umbonula arctica* (M. Sars, 1851)

14b Aperture generally much smaller than half the length of the zooecium; shape of zooecia usually rather constant, the length regularly greater than the width; ooecia present *Stephanosella biaperta* (Michelin, 1845)

15a Avicularium on the midline, at the summit of an eminence; ooecia without pores *Porella*

15b Avicularium to 1 side of the midline, usually on the slope of an eminence (this may be asymmetrical); ooecia with pores *Rhamphostomella*

16a An appreciable part of the frontal surface proximal to the aperture without pores (the area without pores is usually around the midline) 17

16b Pores almost evenly scattered over the frontal surface proximal to the aperture 19

17a Avicularium (typically only 1) on the frontal wall not lateral to the aperture and not adjacent to the proximal margin of the aperture (usually about as far away from the aperture as its own greatest diameter) *Codonellina cribriformis* O'Donoghue & O'Donoghue, 1923

17b With 1 or 2 avicularia either lateral to the aperture or very close to the proximal margin of the aperture 18

18a With 1 or 2 avicularia, these generally lateral to the aperture, but when there is only 1 it may be nearly on the midline close to the proximal margin of the aperture; avicularia not typically on elevations; frontal wall rather granular *Hippomonavella longirostrata* (Hincks, 1883)

18b Usually with 2 avicularia, 1 on each side of the aperture; avicularia usually on distinct elevations (these may be conspicuous and almost tubular); frontal wall not granular, but it may be roughened after calcification is advanced *Costazia*

19a Frontal wall of zooecium with an ascopore (an opening to the compensation sac; it is situated on the midline, close to the proximal margin of the aperture, and must not be confused with an avicularium which may be located in this position in certain genera) 20

19b Frontal wall of zooecium without an ascopore 21

20a With an avicularium on both sides of the aperture, at about the level of the proximal margin of the latter; pores in the frontal wall absent in the area between the ascopore and aperture *Microporella*

20b Avicularia absent altogether; pores in the frontal wall present in the area between the ascopore and the aperture *Fenestratulina malusii* (Andouin, 1826)

21a Avicularia present (although not necessarily throughout the colony), borne on smaller zooecia interspersed among the typical zooecia 22

21b Avicularia present or absent, but if present, found on the frontal surface of typical zooecia, at the sides of the aperture or just proximal to the aperture 23

22a Frontal wall of zooecium with a rimlike thickening at its margins; aperture almost semicircular, the proximal margin nearly straight; zooecia bearing avicularia generally situated between the distal end of 1 zooecium and the proximal end of the succeeding one *Micropora coriacea* (Esper, 1791)

22b Frontal wall of zooecium without a rimlike thickening at its margins; outline of the aperture almost like that of a pear, usually wider in its distal half than in its proximal half, and with a distinct tooth (cardelle) on each side of the proximal half; zooecia bearing avicularia generally situated between adjacent series of typical zooecia *Trypostega claviculata* (Hincks, 1884)

23a Avicularia present 24

23b Avicularia absent 27

24a With an avicularium on both sides of the aperture 25

24b With an avicularium on the midline, rather near to, or actually on, the proximal margin of the aperture 26

25a Apertures on high, tubular elevations, these directed sharply away from the surface of the colony (the elevations are secondary, and as their height increases, the 2 avicularia associated with each zooid form usually distinct tubes alongside them) *Lagenipora*

25b Apertures not on tubular elevations *Schizoporella*

26a Avicularium on the proximal margin of the aperture *Smittina*

26b Avicularium not actually on the proximal margin of the aperture *Schizomavella*

27a Aperture with a single sharp tooth, distally directed, on the proximal lip of the aperture; ooecia not evident at the surface of the colony *Cheilopora praelonga* Hincks, 1883

27b Aperture without a single sharp tooth at the proximal lip of the aperture; ooecia evident at the surface of the colony 27

28a Aperture nearly semicircular, the proximal margin almost straight except for a deep median notch (sinus) (this notch is usually about twice as deep as wide) *Arthropoma cecilii* (Andouin, 1826)

28b Aperture not nearly semicircular, the proximal margin not almost straight, either without a notch or with a wide notch 29

29a Aperture almost circular, without a notch; ooecium often with nearly radial striations *Hippodiplosia insculpta* (Hincks, 1882)

29b Aperture with a wide notch on its proximal margin (unless this is obliterated as the zooecium becomes heavily calcified); ooecium often with tubercles, but without nearly radial striations *Stomachetosella*

30a Frontal wall without transverse ridges and grooves; aperture typically rounded distally, becoming abruptly wider close to its nearly straight proximal margin (the outline of the aperture thus resembles the silhouette of a narrow-brimmed derby hat)
Eurystomella bilabiata (Hincks, 1884)

30b Frontal wall with several prominent transverse ridges alternating with grooves (these represent lines of progressive calcification); aperture, on the whole, generally approximately circular or ovate, with a shallow excavation (sinus) along its proximal margin, but typically becoming abruptly wider at the level of its proximal margin
Hippothoa hyalina (Linnaeus, 1758)

31a Frontal wall of the zooecium decidedly calcified in the peripheral areas, especially in the proximal half of the zooecium, the calcified portion (gymnocyst) generally occupying at least one-fourth (and sometimes more than one-half) of the frontal area 32

31b Frontal wall of the zooecium not extensively calcified (if there is a calcified portion it generally occupies less than one-fourth of the total frontal area) 36

32a Without avicularia 33

32b With avicularia 34

33a With a single, stout, anteriorly-directed spine originating from near the edge of the proximal portion of the calcified area; ooecia absent
Electra crustulenta (Pallas, 1766), var. *arctica* Borg, 1931

33b Without a spine originating near the edge of the proximal portion of the calcified area, but often with a small spine on each side of the operculum (these spines may be insignificant or absent on many of the zooecia); ooecia prominent
Alderina brevispina (O'Donoghue & O'Donoghue, 1926)

34a Spines on the zooecia not branched; ooecia frequently surmounted by an avicularium originating on the proximal portion of the succeeding zooecium (but avicularia may also be lateral to the frontal area)
Cauloramphus, Tegella

(Differentiation of these genera involves fine details; refer to Osburn, 1952.)

34b One or more spines on at least many of the zooecia branched and antlerlike; ooecia not surmounted by an avicularium 35

35a With a small avicularium situated to 1 side of the midline, just proximal to the exposed area of the frontal wall
Doryporella alcicornis (O'Donoghue & O'Donoghue, 1923)

35b Avicularia large, situated between the zooecia, and with a long, slender mandible that fits into a groove in the rostrum
Copidozoum protectum (Hincks, 1884)

36a Without avicularia 37

36b With avicularia 40

37a Peripheral portions of the frontal wall calcified, forming what resembles a narrow shelf extending inward all the way around the zooecium; with a nearly triangular area in each of the proximal corners of many of the zooecia (these areas, if present, at first have a membranous covering, but eventually they become calcified, and may become knoblike and even fused together); on shells and rocks
Conopeum reticulum (Linnaeus, 1767)

37b Peripheral portions of the frontal wall not calcified (however, the lateral walls themselves may be thick, or there may be a delicate calcareous shelf [cryptocyst] extending inward beneath the frontal membrane on the proximal two-thirds of the zooecium); without triangular areas in the proximal corners of the zooecia (but there may be knobs or swellings in the distal corners); regularly on the blades of large kelps, forming silvery patches that may be almost perfectly circular, especially when small 38

38a With a number of small chitinous spines on the frontal membrane and at the margins of the frontal area *Membranipora villosa* Hincks, 1880

38b Without chitinous spines on the frontal membrane or margins of the frontal area 39

39a With a narrow calcareous shelf (cryptocyst) beneath the frontal membrane in the proximal two-thirds of the zooecium *Membranipora serrilamella* Osburn, 1950

39b Without a calcareous shelf beneath the frontal membrane in the proximal two-thirds of the zooecium *Membranipora membranacea* (Linnaeus, 1767)

40a Avicularia stalked, a few on most of the zooecia, situated on the thick lateral walls just outside the series of spines that border the frontal area *Cauloramphus*

40b Avicularia not stalked, situated between the typical zooecia 41

41a Frontal area of the zooecium bordered by spines *Hincksina pallida* (Hincks, 1884)

41b Frontal area of the zooecium not bordered by spines 42

42a Avicularia relatively small, their greatest diameter only about one-fourth that of the zooecia *Ellisina levata* (Hincks, 1882)

42b Avicularia relatively large, their greatest diameter regularly at least half that of the zooecia *Hincksina* (species other than *H. pallida* [41a])

43a Colony as a whole highly calcified and inflexible 44

43b Colony as a whole not so highly calcified as to be inflexible 47

44a Colony like a ruffled, lacy network, with distinct fenestrations *Phidolopora pacifica* (Robertson, 1908)

44b Colony of some other form than a lacy network 45

45a Colony in the form of rough nodules with erect frills, and sometimes also with cylindrical offshoots, otherwise encrusting *Holoporella*

45b Colony decidedly branched (but a sizable basal portion may be encrusting, and often is nodular), the branches more or less cylindrical 46

46a Zooecia not projecting above the surface, the branches therefore nearly smooth *Myriozoum*

46b Zooecia projecting above the surface, the branches therefore rather rough *Costazia*

47a Colony treelike or bushy, generally extensively branched 48

47b Colony in the form of a sheet that is fanlike, leaflike, or lichenlike, or divided into broad, flat branches 56

48a Branches cylindrical, smooth, so conspicuously divided into internodes that the constricted articulations can be observed with the unaided eye 49

48b Branches not cylindrical or smooth (the zooecia are raised), and not so conspicuously divided into internodes that the articulations can be observed with the unaided eye (in *Tricellaria* and *Scrupocellaria*, the branches do have constrictions at rather regular intervals, and in *Caulibugula* there is usually a somewhat articulated main stem, but these details are not obvious unless specimens are examined with a microscope 50

49a Internodes about the shape of elongated cucumbers, and mostly about 1 cm long; articulations not dark; zooecia not hexagonal *Microporina borealis* (Busk, 1855)

49b Internodes very slender and mostly about 2 cm long; articulations dark brown or nearly black; most zooecia rather distinctly hexagonal *Cellaria*

50a Colony treelike, with numerous delicate branches arising from the upper part of a slender stalk composed of a single series of generally elongate zooecia (the stalk often appears to be jointed) *Caulibugula*

50b Colony more nearly bushy than treelike, generally branching from near the base 51

51a Branches rather regularly articulated due to the presence of constrictions 52

51b Branches not regularly articulated (however, a few articulations may sometimes be noted in a colony) 53

52a Zooids with vibracula and avicularia on the surface opposite that on which the aperture is situated *Scrupocellaria*

52b Zooids without vibracula and avicularia on the surface opposite that on which the aperture is situated *Tricellaria*

53a Vibracula numerous, crowded, filling the space on the side of each branch opposite that on which the feeding zooids are located (the setae of the vibracula are so large that the branches appear bristly even when observed with the unaided eye); color generally yellowish or yellowish brown; strictly subtidal 54

53b Vibracula absent; includes some species found intertidally and on floats 55

54a Frontal wall guarded by a spine that is flattened into a bilobed shieldlike or macelike structure *Caberea boryi* (Andouin, 1826)

54b Frontal wall not guarded by a flattened spine *Caberea ellisi* (Fleming, 1828)

55a Zooecia arranged in a single row; zooecia without spines; strictly subtidal and rare *Corynoporella spinosa* Robertson, 1905

55b Zooecia generally in double rows, only occasionally in single rows or several rows; zooecia with 1 or more spines at the distal end, or at least with a spinelike process at the outer distal corner; common intertidally, on floats, and subtidally *Bugula*

56a Ooecia conspicuous, distal to the apertures of the zooecia, generally about one-third the size of the zooecia; spines present at the sides of the zooecia, as well as at the distal corners; avicularia, if present, beaked and on short stalks *Dendrobeania*

56b Ooecia within the zooecia, not visible at the surface; spines, if present at all, small and restricted to the distal corners of the zooecia; avicularia not stalked, oval to nearly quadrangular in outline *Terminoflustra membranacea-truncata* (Smitt, 1867)

Kli͡uge, G. A. 1962. [Bryozoans of the northern seas of the U.S.S.R.] Opredeliteli po Faune SSSR, 76. 584 pp. (in Russian)

McCain, K. W., & J. R. P. Ross. 1974. Annotated faunal list of cheilostome Ectoprocta of Washington state. Northwest Sci., 48:9-16.

O'Donoghue, C. H. 1925. Notes on certain Bryozoa in the collection of the University of Washington. Publ. Puget Sound. Biol. Sta., 5:15-23.

O'Donoghue, C. H., & E. O'Donoghue. 1923. A preliminary list of the Bryozoa (Polyzoa) from the Vancouver Island region. Contr. Canad. Biol.

Fish., n. s., 1:143-201.
------. 1925. List of Bryozoa from the vicinity of Puget Sound. Publ. Puget Sound Biol. Sta., 5:91-108.
------. 1926. A second list of the Bryozoa (Polyzoa) from the Vancouver Island region. Contr. Canad. Biol. Fish., n. s., 3:49-131.
Osburn, R. C. 1950. Bryozoa of the Pacific coast of America. Part 1, Cheilostomata-Anasca. Allan Hancock Found. Pacific Exped., 14:1-269.
------. 1952. Bryozoa of the Pacific coast of America. Part 2, Cheilostomata-Ascophora. Allan Hancock Found. Pacific Exped., 14:271-611.
------. 1953. Bryozoa of the Pacific coast of America. Part 3, Cyclostomata, Ctenostomata, Entoprocta, and addenda. Allan Hancock Found. Pacific Exped., 14:613-841.
Robertson, A. 1905. Non-incrusting chilostomatous Bryozoa of the west coast of North America. Univ. Calif. Publ. Zool., 2:235-322.
------. 1908. The incrusting chilostomatous Bryozoa of the west coast of North America. Univ. Calif. Publ. Zool., 4:253-344.
------. 1910. The cyclostomatous Bryozoa of the west coast of North America. Univ. Calif. Publ. Zool., 6:225-84.
Ross, J. R. P. 1970. Keys to the recent cyclostome Ectoprocta of marine waters of northwest Washington State. Northwest Sci., 44:154-69.

PHYLUM BRACHIOPODA

The key includes only those species that have been found intertidally or that are regularly taken by dredging at moderate depths. For several other species occurring in deep water, consult the paper of Bernard (1972).

The only species that is at all common intertidally is *Terebratalia transversa*. It is extremely variable in shape and in the extent to which the valves are sculptured. Some specimens are nearly smooth; others are strongly ribbed.

1a Shell black or dark gray; posterior beak of ventral (larger) valve rather sharply pointed, its apex not obliterated by the opening through which the attachment stalk passes
Hemithyris psittacea (Gmelin, 1790)

1b Shell yellowish, brownish, or reddish; posterior beak of the ventral (larger) valve not pointed, and its apex may be at least partially obliterated by the opening through which the attachment stalk passes 2

2a Shell usually broader than long; margins of valves at the anterior (broader) end with pronounced complementary undulations
Terebratalia transversa (Sowerby, 1846)

2b Shell usually longer than broad; margins of valves at anterior (broader) end without pronounced complementary undulations (though they may be not quite straight and may be toothed) 3

3a Anterior margins of valves with regularly spaced fine teeth; shell not translucent; valves with fine radiating striations, but usually without concentric growth lines; apex of the ventral valve completely obliterated by the opening through which the attachment stalk passes
Terebratulina unguicula Carpenter, 1845

3b Anterior margins of valves without regularly spaced fine teeth; shell somewhat translucent; valves with crowded concentric growth lines, but without radiating striations; apex of the ventral valve not completely

obliterated by the opening through which the attachment stalk passes
Laqueus californianus (Koch, 1848)

Bernard, F. R. 1972. The living Brachiopoda of British Columbia. Syesis, 5:73-82.

PHYLUM ECHINODERMATA

CLASS CRINOIDEA

ORDER COMATULIDA

Florometra serratissima (A. H. Clark, 1907) is the only crinoid likely to be encountered in our region. It is taken by dredging at depths ranging from about 30 m to over 1000 m, but it is rather local in distribution.

Clark, A. H. 1907. Descriptions of new species of recent unstalked crinoids from the North Pacific Ocean. Proc. U. S. Nat. Mus., 33:69-84.
Clark, A. H., & A. McG. Clark. 1967. A monograph of the existing crinoids. Vol. 1. The Comatulids. Part 5. Suborders Oligophreata (concluded) and Macrophreata. Bull. 82, U. S. Nat. Mus. xiv + 860 pp.

CLASS OPHIUROIDEA

Approximately 25 species of ophiuroids have been reported from this region. However, only about half of these are likely to be encountered intertidally or by dredging at depths no greater than 200 m. The species restricted to deep water are omitted from the key provided here. They can be identified with the help of the substantially complete conspectus of Kyte (1969).

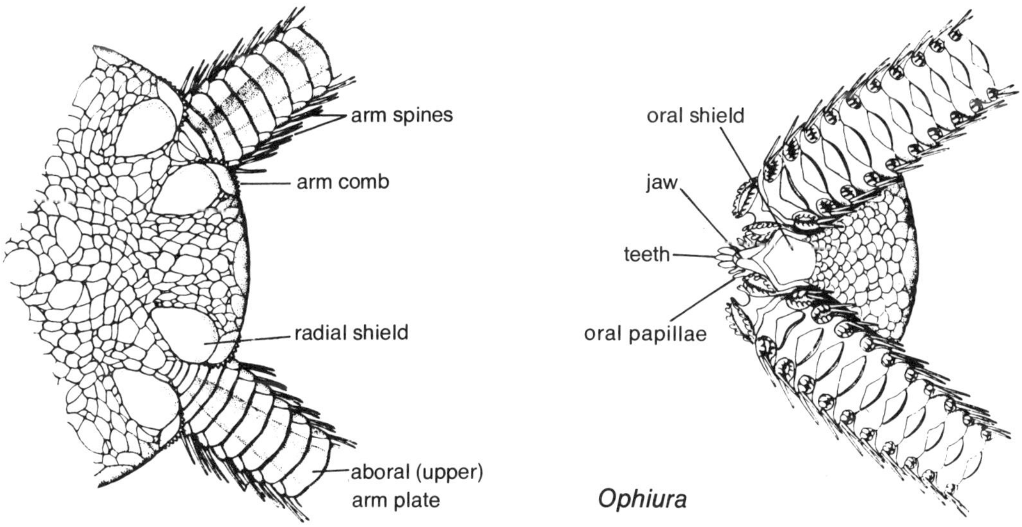

Ophiura

1a Arms repeatedly dichotomously branched; plates of arms and disk concealed by a thick skin; diameter of disk up to about 10 cm; strictly subtidal *Gorgonocephalus eucnemis* (Müller & Troschel, 1842)

1b Arms not branched; plates of arms and disk usually distinct, although those on the disk may be very small or partially obscured by spines; diameter of disk not exceeding 2 cm; intertidal or subtidal 2

2a Plates on upper surface of arms separated from one another by small supplementary plates; coloration usually with some bright red or purplish red, this typically concentrated in streaks or blotches *Ophiopholis aculeata* (Linnaeus, 1767)

2b Plates on upper surface of arms not separated from one another by small supplementary plates; coloration usually not including bright red or purplish red streaks or blotches (however, in *Ophiura leptoctenia,* which is strictly subtidal, the aboral surface of the disk may be rose or red, with a few streaks, and some other species may be pinkish) 3

3a Aboral surface of disk with conspicuous marginal notches into which the arms are inserted, each notch bordered on both sides by an arm comb 4

3b Aboral surface of disk without marginal notches into which the arms are inserted, and without arm combs 6

4a Oral shields (the largest plates on the oral surface, situated between the bases of the arms) widest on the side nearer the margin of the disk; papillae of arm combs at least 5 times as long as wide; prevailing color of aboral surface of disk rose, violet, or red *Ophiura leptoctenia* H. L. Clark, 1911

4b Oral shields widest near their middle portions; papillae of arm combs usually not more than 3 times as long as wide; prevailing color of aboral surface of disk generally some shade of gray, bluish gray, or greenish gray, rarely reddish 5

5a Papillae of arm combs truncate and contiguous *Ophiura lütkeni* (Lyman, 1860)

5b Papillae of arm combs pointed and well separated *Ophiura sarsii* Lütken, 1854

6a With 4 oral papillae on both sides of each jaw 7

6b With 3 oral papillae on both sides of each jaw 8

7a Plates on upper surface of arms nearly circular, hardly wider than long, and barely in contact with one another even near the bases of the arms; interbrachial areas of oral surface of disk (the areas between the arm insertions) without scales *Unioplus strongyloplax* H. L. Clark, 1911

7b Plates on upper surface of arms nearly triangular (but with truncated corners), decidedly wider than long and in contact with one another at least near the bases of the arms; interbrachial areas of oral surface of disk with fine scales *Unioplus macraspis* (H. L. Clark, 1911)

8a Oral papillae of the pair nearest the tip of the jaw decidedly longer than those of the other 2 pairs 9

8b All 3 pairs of oral papillae about the same length 10

9a Length of arms not over 4 and one-half times the diameter of the disk; longest arm spines usually shorter than the width of the arm segments from which they originate; ovoviviparous (the young develop in the bursal pockets) *Amphipholis squamata* (Delle Chiaje, 1828)

9b Length of arms 7 or 8 times the diameter of the disk; longest arm spines longer than the width of the arm segments from which they originate; oviparous (discharging eggs instead of brooding young in the bursal pockets) *Amphipholis pugetana* (Lyman, 1860)

10a Arm spines blunt, strongly flattened *Diamphiodia occidentalis* (Lyman, 1860)

10b Arm spines pointed, either not flattened or only slightly flattened (oval in cross-section) 11

11a Arm spines not at all flattened (circular in cross-section)
Amphiodia urtica (Lyman, 1860)

11b Arm spines slightly flattened (oval in cross-section)
Diamphiodia periercta (H. L. Clark, 1911)

Clark, H. L. 1911. North Pacific ophiurans in the collection of the United States National Museum. Bull. 75, U. S. Nat. Mus. xvi + 302 pp.

D'i͡akonov, A. M. 1967. [Ophiuroids of seas of the U.S.S.R.] Opredeliteli po Faune SSSR, 55. 136 pp. (in Russian)

Kyte, M. A. 1969. A synopsis and key to the recent Ophiuroidea of Washington State and southern British Columbia. J. Fish. Res. Bd. Canada, 26:1727-41.

CLASS ASTEROIDEA

The key includes all sea stars likely to be found intertidally and subtidally to depths of about 200 m. For deep-water species that do not seem to fit, consult the monograph of Fisher (1911-1930).

1a With 5 or 6 rays (except in the case of specimens that have been injured and have regenerated extra rays) 2

1b With 8 to 20 or more rays 15

2a With 6 rays (occasionally 5); maximum diameter about 8 cm
Leptasterias hexactis (Stimpson, 1862)

2b With 5 rays; diameter regularly exceeding 8 cm 3

3a Aboral surface smooth and slippery, with few if any exposed spines 4

3b Aboral surface rough or gritty due to exposed calcareous spines or plates 5

4a Thickness equal to about one-third the diameter; aboral surface with a conspicuous and slightly elevated central opening (this communicates with a spongy cavity that lies between the body wall proper and a superficial membrane); madreporite not evident; color of aboral surface usually pale orange or brownish yellow; without an odor like that of exploded gunpowder, but generally secreting a large amount of nearly gelatinous mucus when first collected; subtidal
Pteraster tesselatus Ives, 1888

4b Thickness less than one-third the diameter; aboral surface without a conspicuous and elevated central opening (there is no spongy cavity covered by a membrane distinct from that of the body wall); madreporite evident, though sunken; color of aboral surface usually reddish brown, with some patches of gray or purple; with an odor like that of exploded gunpowder, but not secreting a large amount of nearly gelatinous mucus when collected; intertidal and subtidal
Dermasterias imbricata (Grube, 1857)

5a Diameter of disk about one-third to one-half the total diameter 6

5b Diameter of disk decidedly less than one-third the total diameter 9

6a Margins of rays and aboral surface studded with large, well-separated, blunt spines *Hippasteria spinosa* Verrill, 1909

6b Margins of rays and aboral surface not studded with large, well-separated, blunt spines 7

7a Margins of rays bordered by large plates that can be seen when the animal is viewed from above; spines of aboral surface arranged in very neat, flat-topped paxillae; subtidal 8

7b Margins of rays not bordered by large plates that can be seen when the

animal is viewed from above; spines of aboral surface not arranged in flat-topped paxillae; intertidal and subtidal, but in our area found only on the open coast of Vancouver Island
Patiria miniata (Brandt, 1835)

8a Outermost small spines constituting the paxillae usually arranged in such a way that they form a very definite peripheral series; total diameter not often exceeding 16 cm; without the odor of exploded gunpowder *Mediaster aequalis* Stimpson, 1857

8b Outermost small spines constituting the paxillae not usually arranged to form a very definite peripheral series; total diameter sometimes exceeding 20 cm; with the odor of exploded gunpowder
Gephyreaster swifti (Fisher, 1905)

9a Pedicellariae absent; aboral surface not noticeably spiny (the small spines are concentrated in paxillae) 10

9b Pedicellariae present; aboral surface noticeably spiny (though the spines nay be blunt) 11

10a Margins of rays with stout spines (these arise from marginal plates); arms decidedly flattened; color of aboral surface generally gray or brownish gray; total diameter frequently exceeding 30 cm; strictly subtidal *Luidia foliolata* Grube, 1866

10b Margins of rays without stout spines (and without marginal plates); arms almost cylindrical; color of aboral surface generally bright orange-red, but sometimes with a mottled pattern involving various shades of red; total diameter rarely exceeding 15 cm; intertidal and subtidal *Henricia leviuscula* (Stimpson, 1857)

11a Spines on upper portions of aboral surface generally not higher than about 2 mm, typically in irregular clusters or arranged in networklike patterns, though some may be single and some may constitute short rows; majority of the pedicellariae not concentrated in raised, almost perfectly hemispherical cushions, with a spine in the center of each 13

11b Spines on upper portions of the aboral surface regularly about 3 to 5 mm long, typically well separated and more inclined to form poorly defined rows than clusters or networks; majority of the pedicellariae concentrated in raised, almost perfectly hemispherical cushions, with a spine in the center of each 12

12a Prevailing coloration of aboral surface some shade of red, often concentrated in blotches or bands contrasting with a light background; teeth on jaws of crossed pedicellariae essentially a series of small serrations, none of them fanglike; intertidal and subtidal
Orthasterias koehleri (de Loriol, 1897)

12b Prevailing coloration of aboral surface dark brown, olive, or gray, without any obvious reddish tones; teeth on jaws of crossed pedicellariae long, the pair at the tip of each jaw especially fanglike; subtidal *Stylasterias forreri* (de Loriol, 1887)

13a Diameter of oral disk not more than one-sixth the total diameter; rays typically broadest a short distance away from the disk; jaws of the straight pedicellariae not forked at their tips into a pair of large blades; color extremely variable, may be gray, greenish, brown, orange, or red, but not likely to be purple or pink
Evasterias troschelii (Stimpson, 1862)

13b Diameter of disk usually greater than one-fifth the total diameter; rays typically broadest where they join the disk; both jaws of the straight pedicellariae forked at their tips into a pair of unequal large blades; color generally orange, orange-ochre, brown, purple, or some shade of pink 14

14a Spines on upper portions of the aboral surface usually arranged in a networklike pattern; color either orange, orange-ochre, brown, or purple; diameter not often exceeding 25 cm; intertidal and subtidal *Pisaster ochraceus* (Brandt, 1835)

14b Spines on upper portions of the aboral surface usually single or clustered, but sometimes arranged in a networklike pattern; color some shade of pink; diameter regularly exceeding 25 cm, and sometimes exceeding 40 cm; subtidal *Pisaster brevispinus* (Stimpson, 1857)

15a With 8 to 16 rays (except in very small individuals) 16

15b With 20 to 24 rays (except in very small individuals) *Pycnopodia helianthoides* (Brandt, 1835)

16a Paxillae of aboral surface rather widely separated and giving rise to long, slender spines of uneven length (thus the aboral surface appears spiny) *Crossaster papposus* (Linnaeus, 1767)

16b Paxillae of aboral surface crowded, giving rise only to short, blunt projections (thus the aboral surface appears rather smooth) 17

17a Diameter of disk usually about one-fourth the total diameter; aboral surface usually some shade of orange or pink, with a grayish blue streak radiating from the center of the disk to the tip of each ray; intertidal and subtidal *Solaster stimpsoni* Verrill, 1878

17b Diameter of disk usually about one-third the total diameter; aboral surface usually some shade of gray or pale brown, without any conspicuous darker streaks radiating from the center of the disk 18

18a Paxillae of the aboral surface almost contiguous; usually with 8 to 11 rays *Solaster endeca* (Linnaeus, 1771)

18b Paxillae of the aboral surface separated from one another to about the extent of their diameter; usually with 12 or 16 rays *Solaster dawsoni* Verrill, 1878

Chia, F. S. 1966a. Systematics of the six-rayed sea star *Leptasterias* in the vicinity of San Juan Island, Washington. Systematic Zool., 15:300-306.

Fisher, W. K. 1911-1930. Asteroidea of the north Pacific and adjacent waters. Bull. 76, U. S. Nat. Mus. Part 1 (1911):vi + 419 pp.; Part 2 (1928):iii + 245 pp.; Part 3 (1930):iii + 356 pp.

CLASS ECHINOIDEA

1a Body discoid, the diameter several times greater than the thickness (Order Clypeastroida--sand dollars) *Dendraster excentricus* (Eschscholtz, 1831)

1b Body not discoid, the diameter never more than twice the thickness 2

2a Body nearly heart-shaped when viewed from the oral or aboral aspect; ambulacra in furrows, one of which is much deeper than the others; symmetry secondarily bilateral, with the mouth lying on the oral surface close to the deepest ambulacral furrow and the anus lying on the margin of the test at the opposite side of the body (Order Spatangoida --heart urchins) *Brisaster latifrons* (A. Agassiz, 1898)

2b Body biscuit-shaped and circular in outline when viewed from the oral or aboral aspect; ambulacra not in well marked furrows; symmetry almost perfectly radial, with the mouth in the center of the oral surface and the anus in the center of the aboral surface (Order Echinoida

--sea urchins) 3

3a Prevailing general coloration greenish or whitish, definitely not red, purple, or orange-pink 4

3b Prevailing coloration red, purple, or orange-pink 5

4a Prevailing coloration of spines whitish, although some, especially near the center of the aboral surface, may be greenish; tube feet usually about the same color as the spines, rarely much darker; diameter of test rarely exceeding 6 cm; strictly subtidal, and usually at depths greater than 50 m
Strongylocentrotus pallidus (G. O. Sars, 1871)

4b Prevailing coloration of spines pale greenish; tube feet darker than spines, and usually purple; diameter of test up to about 8 cm; intertidal as well as subtidal
Strongylocentrotus droebachiensis (O. F. Müller, 1776)

5a Prevailing coloration of living animal orange-pink; cleaned test pale orange-pink, extremely fragile; subtidal at depths greater than 100 m
Allocentrotus fragilis (Jackson, 1912)

5b Prevailing coloration of living animal red or purple; cleaned test usually gray or pale purple, not extremely fragile; intertidal as well as subtidal 6

6a Prevailing coloration usually bright red, reddish-purple, or maroon, although the larger spines of lighter specimens may be rose and the smaller spines almost white; diameter of test up to 15 cm; spines up to 7 cm long in larger individuals; intertidal and subtidal
Strongylocentrotus franciscanus (A. Agassiz, 1863)

6b Prevailing coloration purple; diameter of test not exceeding 9 cm; spines rarely more than 2.5 cm long; largely intertidal on rocky shores having considerable wave action, but subtidal to some extent
Strongylocentrotus purpuratus (Stimpson, 1857)

McCauley, J. E., & A. G. Carey, Jr. 1967. Echinoidea of Oregon. J. Fish. Res. Bd. Canada, 24:1385-1401.

Mortensen, T. 1928-51. A monograph of the Echinoidea. 5 vols. København: C. A. Reitzel.

Swan, E. F. 1953. The Strongylocentrotidae (Echinoidea) of the northeast Pacific. Evolution, 7:269-73.

CLASS HOLOTHUROIDEA

1a Upper surface of body covered with large, fleshy, pointed papillae; oral tentacles moplike; up to about 50 cm long (Order Aspidochirotida)
Parastichopus californicus (Stimpson, 1857)

1b Upper surface of body without large, fleshy, pointed papillae; oral tentacles, if branched, bushy or featherlike, but not moplike; rarely over 30 cm long 2

2a Without tube feet other than the oral tentacles; oral tentacles pinnately branched or unbranched 3

2b With tube feet on at least part of the general body surface; oral tentacles bushy (Order Dendrochirotida) 5

3a Oral tentacles fingerlike, unbranched; body wall thick, without anchorlike spicules that project externally; with a tapering "tail" (Order Molpadida) 4

3b Oral tentacles pinnately branched; body wall thin and nearly translucent, with anchorlike spicules that project externally (these are microscopic, but engage the skin when the animal is handled); slender,

but without a tapering "tail" (Order Apodida)
Leptosynapta clarki Heding, 1928
(*L. clarki*, up to about 10 cm long and with 12 tentacles, is the common intertidal and subtidal species in our area. Heding also described *L. transgressor* and *L. roxtona* from this region, but his material of the former was scanty, and that of the latter was fragmentary. It appears, however, that *L. transgressor* is a valid species, characterized in part by being much smaller than *L. clarki*, and in having fewer [9 to 11] tentacles.)

4a Body rather elongate, with a long "tail;" general coloration usually pinkish, purplish, or silvery gray; intertidal and subtidal, in rather clean sand *Paracaudina chilensis* (J. Müller, 1850)

4b Body stout, with a short "tail;" general coloration red-purple; subtidal, in mud *Molpadia intermedia* (Ludwig, 1894)

5a Inferior surface of body a flattened "sole" with 3 double rows of tube feet; superior surface with hard calcareous plates evident externally 6

5b Inferior surface of body not flattened into a "sole;" no hard calcareous plates evident externally 7

6a Body up to about 8 cm long, with a rounded profile; general coloration orange-brown; oral tentacles red *Psolus chitonoides* H. L. Clark, 1901

6b Body up to about 3 cm long, very low in profile; general coloration pink or lavender; oral tentacles pinkish
Psolidium bullatum Ohshima, 1915

7a General coloration typically reddish brown (whitish specimens have been reported, however); tentacles orange or orange-red; length up to about 30 cm *Cucumaria miniata* (Brandt, 1835)

7b General coloration not reddish brown; tentacles not orange or orange-red; length less than 15 cm (in most species, much smaller than this) 8

8a Color white, cream, pale orange, pale pink or pale brown, not dark and not conspicuously speckled with dark pigment (*Thyone benti*, whose general coloration ranges from pale orange to pale brown, often has small dots of brown pigment, but these do not constitute conspicuous speckles so much as they contribute to a nearly uniform coloration) 9

8b Color dark or conspicuously speckled with dark pigment 12

9a Body crescentic, sharply tapered at both ends; tube feet small and inconspicuous; length rarely exceeding 6 cm
Pentamera populifera (Stimpson, 1864)
(Two other species of *Pentamera*, *P. trachyplaca* (H. L. Clark, 1924) and *P. lissoplaca* (H. L. Clark, 1924) may be expected. The distinctions between them and *P. populifera* have never been clearly explained. Clark's original descriptions and diagnoses given by Deichmann (1937) for *P. populifera* and *P. lissoplaca* may be consulted. In any case, *P. populifera* is the regularly encountered subtidal species in our area.)

9b Body not crescentic, and not sharply tapered at both ends; tube feet, when fully extended, conspicuous; length regularly exceeding 6 cm 10

10a Tube feet restricted to the ambulacra; all 5 ambulacra distinct; length up to about 10 cm, occasionally greater 11

10b Tube feet not restricted to the ambulacra; only 3 ambulacra distinct; length frequently attaining 15 cm *Thyone benti* Deichmann, 1937

11a Tube feet slender; body wall thin; color nearly pure white; often covered by bits of algae, shell and other foreign material; subtidal

Eupentacta pseudoquinquesemita Deichmann, 1938

11b Tube feet stout; body wall tough; color usually cream, but ranging from white to pale orange or orange-pink; not inclined to be covered by debris; on floats and in rocky intertidal areas, as well as subtidal *Eupentacta quinquesemita* (Selenka, 1867)

12a Background color whitish or yellowish, with purple, black, or brown speckles *Cucumaria piperata* (Stimpson, 1864)

12b Color generally rather uniformly grayish, grayish brown, or blackish (however, albinistic populations are sometimes found) 13

13a With many tube feet between the ambulacra; with 4-holed, button-shaped ossicles in the body wall; up to about 5 cm long; subtidal *Cucumaria lubrica* (H. L. Clark, 1901)

13b With few or no tube feet between the ambulacra; without 4-holed, button-shaped ossicles in the body wall; up to about 2 cm long; intertidal, in beds of *Mytilus californianus* *Cucumaria pseudocurata* Deichmann, 1938

Clark, H. L. 1901. The holothurians of the Pacific coast of North America. Zool. Anz., 24:162-71.

------. 1924. Some holothurians from British Columbia. Canad. Field Nat., 38:54-7.

Deichmann, E. 1937. The Templeton Crocker Expedition. IX. Holothurians from the Gulf of California, the west coast of Lower California, and Clarion Island. Zoologica, 22:161-76.

------. 1938. New holothurians from the western coast of North America, and some remarks on the genus *Caudina*. Proc. New England Zool. Club, 16:103-15.

Heding, S. G. 1928. Synaptidae. Papers from Dr. Th. Mortensen's Pacific Expedition 1914-16, no. 46. Vidensk. Medd. Dansk Naturhist. Foren. Copenhagen, 85:105-323.

PHYLUM CHAETOGNATHA

Only 2 species of chaetognaths, *Sagitta elegans* and *Eukrohnia hamata*, are regularly found in coastal waters of our region. *S. elegans* is by far the more common of these. The key includes 2 other species that have been found in small numbers in some collections of plankton.

1a With 2 completely separate lateral fins on each side of the body 2

1b With a long single fin or 2 incompletely separate fins on each side of the body 3

2a Anterior lateral fins slightly longer than posterior lateral fins, and reaching nearly to the level of the posterior edge of the ventral ganglion *Sagitta decipiens* Fowler, 1905

2b Anterior lateral fins slightly shorter than the posterior lateral fins, and terminating at a distance equal to more than half their length behind the posterior edge of the ventral ganglion *Sagitta elegans* Verrill, 1873

3a With 2 incompletely separate lateral fins on each side of the body; eyes pigmented; anus anterior to the tail septum; without oil droplets in the intestine *Sagitta lyra* Krohn, 1853

3b With a single long lateral fin on each side of the body; eyes pigmented; anus at the tail septum; with conspicuous oil droplets in the intestine *Eukrohnia hamata* Möbius, 1875

Fraser, J. H. 1957. Chaetognatha. *In* Fiches d'Identification du Zooplancton (Conseil Permanent International pour l'Exploration de la Mer), No. 1 (revised).

Lea, H. E. 1955. The chaetognaths of western Canadian coastal waters. J. Fish. Res. Bd. Canada, 12:593-617.

PHYLUM HEMICHORDATA

CLASS ENTEROPNEUSTA

Enteropneusts are occasionally found intertidally or subtidally in our area. One species, *Glossobalanus berkeleyi* Willey, 1931, was described from an incomplete specimen collected near Nanaimo. A species of *Saccoglossus*, characterized by a bright orange proboscis, is abundant intertidally in Willapa Bay, Washington, and also in North Bay at Cape Arago, Oregon. What is probably a different species has been reported from Chuckanut Bay, near Bellingham, Washington. We cannot pin names on these, or on others that occasionally turn up, until a long promised systematic study of Pacific coast enteropneusts is completed.

Willey, A. 1931. *Glossobalanus berkeleyi*, a new enteropneust from the west coast. Trans. Royal Soc. Canada, Sect. V, ser. 3, 24:19-28.

PHYLUM CHORDATA

SUBPHYLUM UROCHORDATA

CLASS LARVACEA*

1a Trunk elongate, more than one-third (usually about one-half) the length of the tail *Fritillaria borealis* Lohmann, 1896

1b Trunk ovoid, less than one-third (usually about one-fourth or one-fifth) the length of the tail 2

2a Profile of dorsal surface of trunk somewhat irregular and with a distinct hump near the mouth; with 2 spindle-shaped subchordal cells on the right side of the notochord near the middle of the tail; present in the plankton throughout the year, but especially common from spring to autumn *Oikopleura dioica* Fol, 1872

2b Profile of dorsal surface of trunk rather smooth and without a distinct hump near the mouth; with about 15 to 25 large cuboidal subchordal cells along the notochord in the terminal third of the tail; present in the plankton from fall to spring, but most common in winter *Oikopleura labradoriensis* Lohmann, 1892

Berrill, N. J. 1950. The Tunicata with an Account of the British Species. London: Ray Society. iii + 354 pp.

Essenberg, C. E. 1926. Copelata from the San Diego region. Univ. Calif. Publ. Zool., 28:399-521.

*The key is based on information kindly supplied by Charles Galt.

Tokioka, T. 1960. Studies on the distribution of appendicularians and some thaliaceans of the North Pacific, with some morphological notes. Publ. Seto Mar. Biol. Lab., 8:351-443.

CLASS ASCIDIACEA

Charles C. Lambert
California State University, Fullerton

1a Zooids solitary (although they may be attached to one another) or social (connected by stolons or by sheets of tunic material), but not embedded in a common tunic 2

1b Zooids (several to many) embedded in a common tunic, which is typically gelatinous in texture (compound ascidians) 28

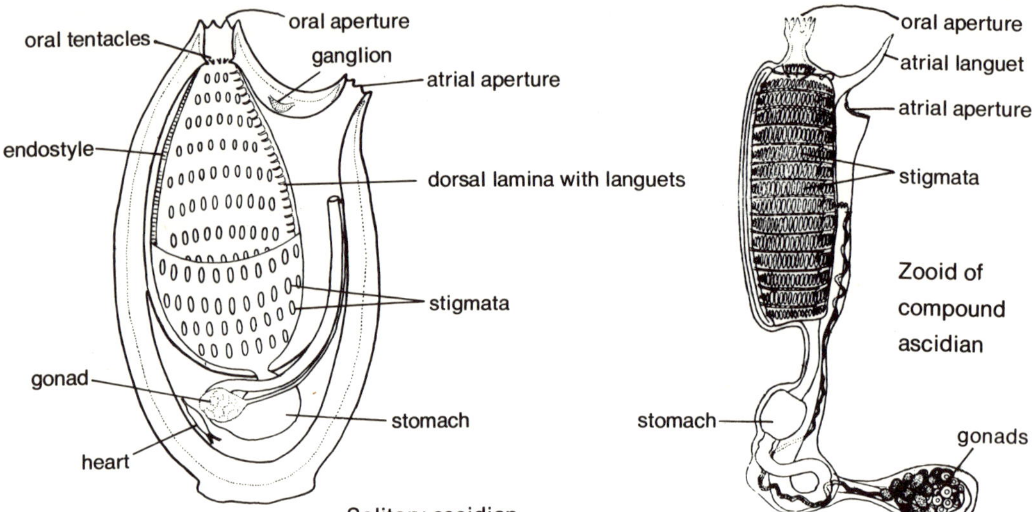

2a Zooids solitary (although they may be attached to one another), not connected by stolons or sheets of tunic material; usually more than 1 cm in diameter (solitary ascidians) 3

2b Zooids social, connected by stolons or by sheets of tunic material (although the connections may not be obvious and may not persist); usually less than 1 cm in diameter (social ascidians) 23

3a Tunic transparent or translucent 4

3b Tunic opaque 11

4a Both the oral and the atrial apertures situated on a flattened disk consisting of a number of distinct plates 5

4b Oral and atrial apertures not situated on a flattened disk consisting of a number of distinct plates 6

5a Plates of disk usually showing concentric growth lines; muscle strands connecting the 2 central plates of the disk not visible through the tunic; no intermediary plates between central and marginal plates; diameter of disk often exceeding 1.5 cm; intertidal, subtidal, and on floats *Chelyosoma productum* Stimpson, 1864

5b Plates of disk not showing concentric growth lines; muscle strands connecting the two central plates of the disk visible through the tunic in preserved specimens; 1 to 3 intermediary plates between central and marginal plates; diameter of disk rarely exceeding 1.5 cm; subtidal *Chelyosoma columbianum* Huntsman, 1912

6a Body not attached to the substrate; tunic covered with fine, short, hairlike projections *Molgula pugetiensis* Herdman, 1898

6b Body attached to a firm substrate; tunic smooth or with papillae, but without hairlike projections 7

7a Body elongate and somewhat compressed 9

7b Body low, nearly hemispherical 8

8a Atrial aperture usually situated near the end of the anterior one-third of the body; with about 25 to 35 oral tentacles (these are located around the opening to the pharynx); common subtidally and on floating docks, and occasionally found in the intertidal zone *Ascidia callosa* Stimpson, 1852

8b Atrial aperture usually situated near the end of the anterior half of the body; with about 150 to 200 oral tentacles; in our area, found only subtidally *Ascidia ceratodes* (Huntsman, 1912)

9a Tunic smooth or only irregularly wrinkled 10

9b Surface of tunic with scattered large papillae *Ascidia paratropa* (Huntsman, 1912)

10a Body not much longer than wide; tunic transparent and colorless; longitudinal muscle bands not obvious beneath the tunic; oral and atrial apertures not borne on tubes *Corella willmeriana* Herdman, 1898

10b Body much longer than wide; tunic transparent, but tinted yellow-green; longitudinal muscle bands distinctly visible beneath the tunic; oral and atrial apertures borne on short tubes ("siphons") *Ciona intestinalis* (Linnaeus, 1767)

11a Tunic covered with spinelike projections 12

11b Tunic smooth or wrinkled, but without spinelike projections 13

12a Body attached to the substrate by a distinct stalk; surface of tunic plainly visible between the short spinelike projections arising from it *Boltenia villosa* (Stimpson, 1864)

12b Body not attached to the substrate by a distinct stalk; surface of tunic concealed by the dense covering of long spinelike projections arising from it *Halocynthia igaboja* Oka, 1906

13a Both the oral and the atrial apertures situated on a flattened disk consisting of a number of distinct plates 14

13b Oral and atrial apertures not situated on a flattened disk consisting of a number of distinct plates 15

14a Plates of disk usually showing concentric growth lines; muscle strands connecting the 2 central plates of the disk not visible through the tunic; no intermediary plates between central and marginal plates; diameter of disk often exceeding 1.5 cm; intertidal, subtidal, and on floats *Chelyosoma productum* Stimpson, 1864

14b Plates of disk not showing concentric growth lines; muscle strands connecting the 2 central plates of the disk visible through the tunic in preserved specimens; 1 to 3 intermediary plates between central and marginal plates; diameter of disk rarely exceeding 1.5 cm; subtidal *Chelyosoma columbianum* Huntsman, 1912

15a Body elongate, attached to the substrate by a rather small area of the tunic 16

15b Body not elongated, attached to the substrate by a rather large area of the tunic 18

16a With a slender, furrowed stalk that is about half the length of the body; restricted to the open coast *Styela montereyensis* (Dall, 1872)

16b Without a slender furrowed stalk; not restricted to the open coast 17

17a Body barrel-shaped; surface of tunic smooth or with only fine furrows and wrinkles *Halocynthia aurantium* (Pallas, 1787)

17b Body elongate, cylindrical, shaped much like a small cucumber; tunic with many longitudinal wrinkles *Styela gibbsii* (Stimpson, 1864)

18a Tunic smooth, glabrous; color of entire animal bright orange-red *Cnemidocarpa finmarkiensis* (Kiaer, 1893)

18b Tunic coarsely or finely wrinkled, never glabrous; bright red color, if present, restricted to areas surrounding oral and atrial apertures 19

19a Body elongated; oral and atrial apertures at opposite ends of the body *Pyura mirabilis* (von Drasche, 1884)

19b Body not elongated; oral and atrial apertures not at opposite ends of the body 20

20a Oral and atrial apertures situated on conspicuous tubes ("siphons"), the oral tube being decidedly longer than the atrial tube; tunic tough, thick, and coarsely wrinkled *Pyura haustor* (Stimpson, 1864)

20b Oral and atrial apertures not situated on conspicuous tubes; tunic stiff, but not coarsely wrinkled 21

21a Attachment to substrate by the left side of body, or partly by the left side and partly by the ventral side; tunic fairly smooth, but often encrusted with foreign material or small organisms; tunic nearly colorless to light brown or greenish 22

21b Attachment to substrate by the ventral side of body; tunic finely wrinkled and studded with small translucent tubercles; tunic dull brown *Styela coriacea* (Alder & Hancock, 1848)

22a Atrial aperture usually situated near the end of the anterior one-third of the body; with about 25 to 35 oral tentacles (these are located around the opening to the pharynx); common subtidally and on floating docks, occasionally found in the intertidal zone *Ascidia callosa* Stimpson, 1852

22b Atrial aperture usually situated near the end of the anterior half of the body; with about 150 to 200 oral tentacles; in our area, found only subtidally *Ascidia ceratodes* (Huntsman, 1912)

23a Individuals not much, if any, taller than wide 24

23b Individuals much taller than wide 26

24a Individuals transparent and green; attached to the substrate by a rather narrow portion of the tunic *Perophora annectens* Ritter, 1893

24b Individuals opaque and usually orange-red or brick red (sometimes yellowish); attached to the substrate by a rather broad portion of the tunic 25

25a Contiguous individuals joined by fine stolons or by thin sheets of tunic material, the connections often indistinct; on rocks, intertidal and subtidal *Metandrocarpa taylori* Hunstman, 1912

25b Contiguous individuals joined to the extent that they appear to constitute a compound ascidian; usually on kelp, subtidal *Metandrocarpa dura* (Ritter, 1896)

26a Individuals widely separated; pharynx yellow-orange; less than 1 cm tall when mature *Pycnoclavella stanleyi* Berrill & Abbott, 1949

26b Individuals clustered; pharynx, if pigmented, not yellow-orange (caution! -- the tadpoles are yellow); usually taller than 1 cm when mature 27

27a Dorsal lamina and endostyle bright fluorescent pink; apertures not lobed *Clavelina huntsmani* Van Name, 1931

27b Dorsal lamina and endostyle not fluorescent pink; oral aperture irregularly lobed, atrial aperture with 2 lips, each in turn divided into 3 small lobes *Clavelina* sp.

28a Colonies forming thin encrusting sheets, thick encrusting slabs (often irregular and lobed), or low mounds attached for most of their diameter to the substrate (the colony may be very lumpy, but it does not consist of club-shaped, mushroomlike, leaflike, or globular masses that have distinct stalks) 29

28b Colonies forming club-shaped, mushroomlike, leaflike, or globular masses, usually with distinct stalks (even when the masses are moundlike, they are attached by a narrow stalk) 35

29a Colony encrusting, not more than 3 mm thick, and not inclined to be lumpy 30

29b Colony fleshy, forming cakes, low mounds, or slabs (these often irregular, lobed, and lumpy) 5 mm to 1 cm or more in thickness 32

30a Colony an opaque whitish or grayish encrustation only about 1 or 2 mm thick; tunic firm, containing blunt-rayed globular spicules (the degree of firmness and whiteness varies according to the density of the spicules) 31

30b Colony transparent, colorless, gray, olive, or tan, up to about 3 mm thick; tunic soft and gelatinous, lacking calcareous spicules (though it may have scattered flecks of white pigment) *Diplosoma macdonaldi* Herdman, 1886

31a Pharynx with 4 rows of stigmata *Didemnum albidum* (Verrill, 1871)

31b Pharynx with 3 rows of stigmata *Trididemnum ?carnulentum* Ritter & Forsyth, 1917

32a Tunic with disklike calcareous spicules and bladderlike cells (both visible at a magnification of 15 times) *Cystodytes lobatus* (Ritter, 1900)

32b Tunic without disklike calcareous spicules or bladderlike cells 33

33a Most or all zooids arranged in "systems;" the individual zooids of a system have their oral apertures at the surface, but their atrial apertures communicate with a cavity ("cloaca") that has a single opening 34

33b Zooids not arranged in "systems;" the oral and atrial apertures of each zooid open at the surface *Archidistoma ritteri* (Van Name, 1945)

34a Zooids less than 1 cm in length, stomach ridges regular and unbranched *Aplidium californicum* (Ritter & Forsyth, 1917)

34b Zooids up to 2 cm in length, stomach ridges irregular and branched *Aplidium ?constellatum* (Verrill, 1871)

35a Pharynx with 4 rows of stigmata; colonies not sand-encrusted 36

35b Pharynx with 5 or more rows of stigmata; colonies often sand-encrusted 37

36a Colony consisting of club-shaped or mushroomlike masses, sometimes broad mounds (but these are attached by narrow stalks); color ranging from pale orange to dark purplish red; found in protected waters, especially on floats and pilings, as well as on the outer coast *Distaplia occidentalis* Bancroft, 1899

36b Colony consisting of a cluster of leaflike or paddlelike lobes; color ranging from cream to light orange-brown; restricted to the outer coast *Distaplia smithi* Abbott & Trason, 1968

37a Zooids arranged in "systems" (the individual zooids in a system have their oral apertures at the surface, but their atrial apertures communicate with a cavity ["cloaca"] that has a single opening) 38

37b Zooids not arranged in systems (both the oral and atrial apertures of each zooid open at the surface)
Ritterella aequalisiphonis (Ritter & Forsyth, 1917)

38a Atrial aperture of each zooid not situated at the end of a distinct tube; atrial languet (a slender, tonguelike outgrowth of the lip of the atrial aperture) present 39

38b Atrial aperture of each zooid situated at the end of a distinct tube; atrial languet absent *Synoicum ?jordani* (Ritter, 1899)

39a Pharynx with 5 rows of stigmata; club-shaped lobes of colony rather slender, containing only a few zooids; tunic both encrusted and impregnated with sand *Aplidium arenatum* (Van Name, 1945)

39b Pharynx with 10 or more rows of stigmata; lobes of colony slender or globular or irregular, containing few to many zooids; tunic, if encrusted with sand, not also impregnated with it 40

40a Pharynx with 10 to 12 rows of stigmata; lobes of colony usually irregularly club-shaped, encrusted with sand only at the base if at all 41

40b Pharynx with 17 or more rows of stigmata; lobes of colony usually globular and often somewhat flattened above, encrusted with sand
Aplidium propinquum (Van Name, 1945)

41a Atrial languet usually with a pair of lateral lobes arising near its base (when these lobes are well developed, the languet is trident-like); restricted to the outer coast *Aplidium glabrum* (Verrill, 1871)

41b Atrial languet without a pair of lateral lobes; found in protected waters as well as on the outer coast
Aplidium ?constellatum (Verrill, 1871)

Abbott, D. P., & W. Trason. 1968. *Ritterella rubra* and *Distaplia smithi*. Bull. So. Calif. Acad. Sci., 67:143-53.

Berrill, N. J. 1950. The Tunicata with an Account of British Species. London: Ray Society. iii + 354 pp.

Huntsman, A. G. 1912. Holosomatous ascidians from the coast of western Canada. Contrib. Canad. Biol., 1906-1910:103-85.

Ritter, W. E. 1900. Some ascidians from Puget Sound, collections of 1896. Ann. New York Acad. Sci., 12:589-616.

------. 1913. The simple ascidians from the northeastern Pacific in the collection of the United States National Museum. Proc. U. S. Nat. Mus., 45:427-505.

Van Name, W. G. 1945. The North and South American ascidians. Bull. Amer. Mus. Nat. Hist., 84:1-476.

CLASS THALIACEA

Family Salpidae

The salps are pelagic animals that are almost completely restricted to the open ocean. They are not often taken in collections of plankton made in inland waters. However, they are occasionally washed up on the shore, especially on the outer coast.

The paper of Yount (1954), concerned with the rich fauna of salps found in the central part of the Pacific Ocean, provides a key to all valid species, a few of which occur in our general area. The other works listed in the bibliography will be helpful for confirming identifications.

Apstein, C. 1901. Salpidae, Salpen. Nordisches Plankton, Zool., 2, Abt.

III:5-10.
Fraser, J. H. 1947. Thaliacea--I. Family: Salpidae. Fiches d'Identification du Zooplancton (Conseil Permanent International pour l'Exploration de la Mer), Sheet 9. 4 pp.
Metcalf, M. M. 1919. The Salpidae collected by the United States Fisheries Steamer "Albatross," in Philippine waters, during the years 1908 and 1909. Bull. 100, U. S. Nat. Mus., vol. 2, part 2. 193 pp.
Yount, J. L. 1954. The Taxonomy of the Salpidae (Tunicata) of the central Pacific Ocean. Pacific Sci., 9:276-330.

Family Doliolidae

Doliolids are rarely taken except in plankton tows made well away from the coast. Any species likely to be encountered in our area can probably be identified with the aid of the references listed.

Borgert, A. Die nordischen Dolioliden. Nordisches Plankton, Zool., 2, Abt. III:1-4.
Fraser, J. H. 1947. Thaliacea--II. Family: Doliolidae. Fiches d'Identification du Zooplancton (Conseil Permanent International pour l'Exploration de la Mer), Sheet 10. 4 pp.

GLOSSARY

Aboral. Opposite the end or side on which the mouth is located

Accessory flagellum. In certain Crustacea, a branch of the antennule

Acicle. A spinelike projection

Aciculum (acicular seta). In polychaete annelids, a chitinous needlelike structure that provides internal support for the lobes of a parapodium

Adductor muscle scar. In shells of bivalve molluscs, the scar of a large muscle that pulls the valves together (there is usually an anterior and a posterior adductor muscle)

Ambulacrum. In echinoderms, a region (often marked by a groove) along which the tube feet (podia) are arranged

Antenna. In Crustacea, the second antenna (antenna 2); in polychaete annelids, a slender appendage arising from the prostomium

Antennal scale. In certain Crustacea, the flattened exopodite of the antenna

Antennule. In Crustacea, the first antenna (antenna 1)

Article. In arthropods, a unit ("segment") of an appendage

Atrial languet. In zooids of compound ascidians, a tonguelike process (sometimes divided) arising from the lip of the atrial aperture

Avicularium. In bryozoans, a specialized individual that has two jaws (in some species, the avicularia resemble beaks of birds)

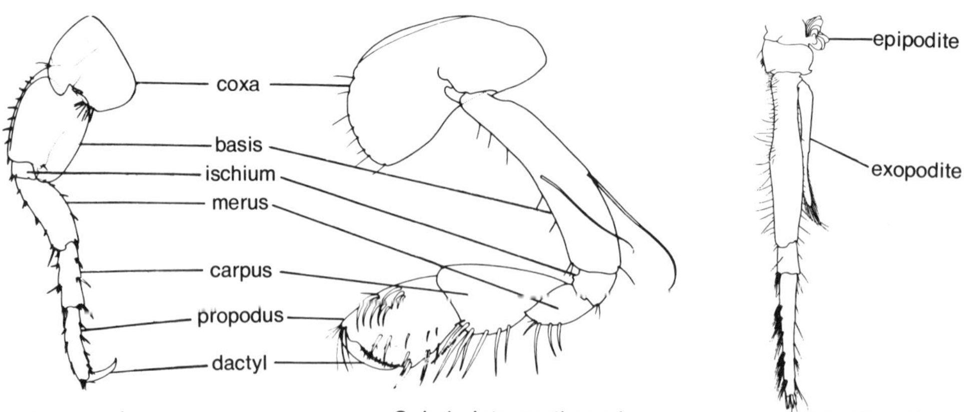

Pereopod　　Subchelate gnathopod　　Maxilliped

Basis (Basopodite). In Crustacea, the second article of an appendage; if the appendage is biramous, the basis is the article that bears the exopodite (see figure)

Body whorl. In the coiled shell of a gastropod mollusc, the youngest (and usually the largest) whorl

Branchia. A structure that functions as a gill

Caecum. A blind pouch; a diverticulum

Cancellate sculpture. In shells of gastropod molluscs, a pattern of sculpture formed by interaction of axial ribs and spiral ridges; in bivalve molluscs, a pattern formed by interaction of radial ribs and concentric ridges or lines

Capillary seta. A very slender, hairlike seta

Carapace. In Crustacea, a covering (generally hardened to at least some extent) over part or all of the body; basically, it is derived as a posterior outgrowth of the last embyronic somite of the head, but it may be fused with outgrowths of other head somites

Cardinal teeth. In the shell of a bivalve mollusc, the teeth that radiate from the part of the hinge immediately adjacent to the umbone

Carpus. In the leg (pereopod) of a crustacean, the fifth article (see p. 208)

Cephalic. Pertaining to the head region

Cephalon. In arthropods, the head, derived from the first six embryonic somites; the term is often applied (as in the case of the isopods) to what is essentially a short cephalothorax, derived from the head plus one thoracic somite

Cerata. In opisthobranch gastropod molluscs, fleshy processes of the dorsum, often containing blood vessels, diverticula of the digestive gland, and nematocysts derived from cnidarians that have been eaten

Chela. In Crustacea, a pincer in which the dactyl serves as a movable claw and the propodus serves as the stationary claw

Chelate. Provided with a chela *(see also* Subchelate)

Cheliped. A leg (pereopod) provided with a chela

Chondrophore. In the shell of a bivalve mollusc, a plate or shelf to which an internal hinge ligament is attached

Chromatophore. A structure (sometimes a single cell) in which pigment is concentrated

Cirrus. In general, a soft appendage, usually fingerlike or tentaclelike; in barnacles, a featherlike thoracic appendage used for collecting food

Colloblast. In ctenophores, a "glue cell," used in capture of prey

Corbula. In certain hydroids, a basketlike structure within which the gonophores are concentrated

Cormidium. In siphonophores, a lateral branch of the stem that trails the swimming bell; it consists initially of a gastrozooid, and this eventually produces one or more gonophores

Coupling hook. On a maxilliped of an isopod crustacean, a small, hooklike projection on the mesial margin

Coxa. In Crustacea, the first article of an appendage; if an epipodite is present, it is borne on the coxa (see p. 208)

Coxal plate. In amphipod Crustacea, a coxa expanded into a broad plate

Crenate, Crenulate. Having a scalloped margin

Cryptocyst. In bryozoans, a calcified shelf beneath the peripheral portion of the frontal membrane

Dactyl. In Crustacea, the terminal article, usually clawlike, of a leg

(pereopod)

Dextral. Referring to the right, or something turned or coiled toward the right

Dorsum. The upper surface of an animal; the back

Dorsal lamina. In ascidians, a narrow membrane running along the mid-dorsal line of the pharynx

Elytron. In certain families of polychaete annelids, a shieldlike plate on the dorsal surface

Endopodite. In Crustacea, the inner (mesial) branch of a biramous appendage

Endostyle. In ascidians, a grooved band of ciliated and glandular tissue running along the mid-ventral line of the pharynx

Epibenthic. Living at the surface of a bottom deposit

Epimeron, epimeral plate. In Crustacea, a flattened lateral extension of a segment, directed laterally or ventrolaterally

Epipodite. In certain Crustacea, a lateral extension of the basis of a leg (pereopod), usually functioning as a gill (see p. 208)

Eudoxid. In siphonophores, a cormidium that has become detached from the stem on which it was produced and that may survive for a time as a free-living individual

Exopodite. In Crustacea, the outer (lateral) branch of a biramous appendage

Falcigerous seta. In polychaete annelids, a compound seta in which the distal portion is shaped something like a sickle

Flagellum. In certain Crustacea, the slender distal portion of an antennule or antenna; the term is more generally used to indicate a vibratile protoplasmic extension of a cell, involved in locomotion, attachment, or creating water currents

Gastrozooid. In Hydrozoa and some other cnidarians, an individual specialized for feeding; a feeding polyp

Girdle. In chitons, the firm mantle that encircles the eight valves, or in which the valves are embedded

Gnathopod. In amphipods and certain other Crustacea, one or both of the first two legs (pereopods), often differing from the remaining legs in being chelate or subchelate

Gonophore. In Hydrozoa generally, a structure within which medusae are produced (the medusae may be abortive and be retained within the gonophore); in siphonophores, the term is generally applied directly to medusiform sexual individuals

Gonopore. A genital pore

Gonotheca. In Hydrozoa, a hardened outer covering (perisarc) enclosing a gonophore

Gymnocyst. In bryozoans, a calcified portion of the frontal wall, proximal and lateral to the membranous portion

Heterogomph. In polychaete annelids, referring to compound setae in which the two lobes at the end of the proximal portion are decidedly unequal

Homogomph. In polychaete annelids, referring to compound setae in which the two lobes at the end of the proximal portion are approximately equal

Hydranth. In Hydrozoa, a feeding polyp; a gastrozooid

Hydrotheca. In hydroids, a hardened covering (perisarc) around a hydranth

Introvert. In sipunculans, a portion of the body that can be retracted by being pulled into itself

Ischium. In Crustacea, the third article of a leg (pereopod) (see p. 208)

Lateral teeth. In the shell of a bivalve mollusc, any teeth that may be present of the hinge plate anterior or posterior to the cardinal teeth

Lithocyst. In hydrozoan medusae, a structure concerned with balance, consisting of a crystalline mass within a chamber containing sensory cells

Lophophore. In bryozoans, brachiopods, and phoronids, a fold or ridge on the body wall that bears ciliated tentacles used in collecting food and in respiration

Madreporite. A perforated calcareous plate ("sieve plate") on the aboral surface of echinoid and asteriod echinoderms, permitting water to enter the water-vascular system (a simplified madreporite is found on the oral surface of some ophiuroids)

Mandible. In Crustacea, one of the third pair of appendages of the head (the first pair of appendages associated with the mouth)

Manubrium. In medusae, a stalk on which the mouth is located

Maxilla. In Crustacea, a member of the fourth and fifth pairs of appendages of the head (maxillae 1, maxillae 2), which follow the mandibles

Maxilliped. In Crustacea, a member of a pair of anterior thoracic appendages that follow the maxillae and that function as mouth parts (the number of pairs of maxillipeds varies according to the group of Crustacea)

Merus. In Crustacea, the fourth article of a leg (pereopod) (see p. 208)

Molar. In Crustacea, a process of a mandible specialized for grinding food

Nectophore. In siphonophores, a swimming bell

Nematocyst. In cnidarians, a stinging capsule, used for capturing prey

Nematophore. In certain hydroids, a polyp specialized for production of nematocysts

Neuroseta. In polychaete annelids, one of the setae arising from the more nearly ventral lobe (neuropodium) of a parapodium

Notoseta. In polychaete annelids, one of the setae arising from the more nearly dorsal lobe (notopodium) of a parapodium

Nuchal. Referring to the neck region

Ocellus. A simple eye; an eyespot

Ommatidium. In arthropods, a unit of a compound eye

Ooecium (ovicell). In bryozoans, a chamber in which the embryo develops

Operculum. A flap or trap door, as used for closing the aperture of a snail shell or the hydrotheca of a hydroid

Osculum. In sponges, an opening by which water leaves the colony

Ossicle. A skeletal structure, usually calcareous, in the form of a plate or spine

Pallial line. In bivalve molluscs, a line that marks the attachment of the mantle to the shell

Pallial sinus. In bivalve molluscs, an indentation of the pallial line, indicating the area into which the siphons can be withdrawn

Palm. In Crustacea, the broadened portion of the propodus (sixth article) of a chelate or subchelate appendage

Papilla. A small, fleshy projection of the body wall or of some other structure.

Paragnath. In certain polychaete annelids (as members of the family Nereidae), one of the small teeth on the outside of the everted proboscis

Parapodium. In polychaete annelids, a fleshy flap on each side of at least certain segments (the setae arise from the parapodia)

Paxilla. In asteroid echinoderms, a columnar calcareous structure on the body surface, the top of which is characteristically nearly flat and covered with small tubercles or blunt spines

Pectinate. Comblike

Pedicellaria. In asteroid and echinoid echinoderms, a pincerlike structure (generally with at least a short stalk) on the body surface (in asteroids, the pedicellariae have two jaws; in echinoids, they usually have three jaws)

Peduncle. A stalk; in the case of Crustacea, applied to the proximal portion of an antennule or antenna, as distinct from the flagellum

Pereon. In Crustacea, the complex of thoracic segments from which the legs (pereopods, gnathopods) originate (it does not include segments that are fused to the head, and whose appendages [maxillipeds] function as mouth parts)

Pereonite. A segment of the pereon

Pereopod. An appendage (leg) arising from a segment of the pereon

Perfoliate. Referring to a structure in which the stalk is encircled by one or more continuous flanges

Periostracum. In molluscs, the organic material, often fibrous, on the outside of the shell

Peristomium. In polychaete annelids, the first true segment, directly behind the prostomium, on which the mouth is located (additional segments may be fused with it, and the complex as a whole may bear tentacular cirri; these are modifications of parapodia, and must not be confused with antennae, which are situated on the prostomium)

Pleon. In Crustacea, the abdomen (not all six segments may be evident, however)

Pleonite. A segment of the pleon

Pleopod. In crustacea, a paired biramous appendage arising from the ventral side of an abdominal segment (pleonite), used in swimming, copulation, tending eggs, and other functions

Pleotelson. In certain crustacea, as isopods, a complex formed by fusion of the telson with one or more segments of the pleon

Propodus. In Crustacea, the sixth (next to last) article of a leg (pereopod, gnathopod) (see p. 208)

Prostomium. In polychaete annelids, the head lobe, dorsal and anterior to the mouth (if antennae or palps are present, these originate from the prostomium)

Protandric. Referring to an animal that is first male, then female

Pseudohydrotheca. In certain hydroids, an extension of the perisarc that encloses at least a part of a hydranth but that differs from a true hydrotheca in being of inconsistent and somewhat irregular form

Rachis. The main stem of a branched structure, especially one that is featherlike

Rhinophore. In opisthobranch gastropod molluscs, a member of a pair of processes, sometimes with a sheath and often elaborate, arising from the dorsal part of the head

Rhopalium. In scyphozoan medusae, a structure concerned with equilibrium, consisting of a fleshy outgrowth of the margin of the bell, weighted by a mass of crystals, that contacts a sensory lobe

Rostrum. In Crustacea generally, an anteriorly directed prolongation of the head or the carapace; in barnacles, the term is applied to one of the plates constituting the armor (see p. 136)

Setigerous. Bearing setae

Setose. Bristly

Sinistral. Referring to something that is left-handed, or turned or coiled toward the left

Siphonoglyph. In Anthozoa, a ciliated groove running the length of the pharynx (there may be one or two of these, and they are usually evident at the level of the mouth)

Stigmata. In ascidians, the perforations of the pharynx

Stolon. A "runner," creeping over or through the substrate, from which new individuals are budded

Stylet. A spinelike or needlelike structure, as on the proboscis of some nemerteans

Subchelate. Descriptive of an appendage, such as a gnathopod of an amphipod crustacean, in which the dactyl closes against the propodus, but in which the propodus does not have a prolongation serving as a stationary claw (see p. 208)

Telson. In Crustacea, a terminal flap or plate attached to the sixth abdominal segment (in certain groups, as the isopods, it may be indistinct from one or more of the abdominal segments [*see* Pleotelson])

Tentacular bulb. In hydrozoan medusae, a swelling at the base of a tentacle

Terete. Approximately circular in cross-section, rather than flattened

Theca. An outer covering, generally hardened to at least some extent

Tracheal system. In insects and certain crustacea, a system of tubes for breathing air

Triturative. Specialized for chewing or grinding

Tunic. The outer covering, often rather thick, of a solitary or social ascidian, and the matrix in which the several to many zooids of a compound ascidian are embedded

Umbilicus. In gastropod molluscs, a pit at the base of the shell, leading into the pillar around which the whorls are spiral

Umbone. In the shell of a bivalve mollusc, the oldest portion of each valve, near the hinge, often elevated or somewhat beaklike

Uropod. In Crustacea, a member of one or more pairs of appendages arising from the posterior part of the abdomen (pleon)

Urosome. In amphipod Crustacea, the complex formed by the last three segments of the abdomen (pleon); these segments bear the three pairs of uropods and the telson

Urosomite. In amphipod Crustacea, one of the last three segments of the abdomen (pleon)

Velum. In hydrozoan medusae, a circular membrane extending inward from the margin of the bell

Vibraculum. In bryozoans, a specialized individual consisting largely of a slender, vibratile projection

Zooecium. In bryozoans, the boxlike or tubelike "house" secreted by a zooid

Zooid. In bryozoans and compound ascidians, an individual member of a colony

INDEX